Atmospheric Chemistry

From the Surface to the Stratosphere

Essential Textbooks in Chemistry

ISSN: 2059-7738

Orbitals: With Applications in Atomic Spectra
by Charles Stuart McCaw

Principles of Nuclear Chemistry
by Peter A C McPherson

Atmospheric Chemistry: From the Surface to the Stratosphere
by Grant Ritchie

Astrochemistry: From the Big Bang to the Present Day
by Claire Vallance

Forthcoming

Problems of Instrumental Analytical Chemistry: A Hands-On Guide
by J M Andrade-Garda, A Carolsena-Zubieta, M P Gómez-Carracedo, M A Maestro-Saavedra, M C Prieto-Blanco and R M Soto-Ferreiro

Essential Textbooks in Chemistry

Atmospheric Chemistry

From the Surface to the Stratosphere

Grant Ritchie

Oxford

W🌐 World Scientific

NEW JERSEY · LONDON · SINGAPORE · BEIJING · SHANGHAI · HONG KONG · TAIPEI · CHENNAI · TOKYO

Published by

World Scientific Publishing Europe Ltd.

57 Shelton Street, Covent Garden, London WC2H 9HE

Head office: 5 Toh Tuck Link, Singapore 596224

USA office: 27 Warren Street, Suite 401-402, Hackensack, NJ 07601

Library of Congress Cataloging-in-Publication Data

Names: Ritchie, G. A. D. (Grant A. D.)

Title: Atmospheric chemistry : from the surface to the stratosphere / Grant Ritchie (Oxford).

Description: New Jersey : World Scientific, 2016. | Series: Essential textbooks in chemistry | Includes index.

Identifiers: LCCN 2016035655| ISBN 9781786341754 (hc : alk. paper) | ISBN 9781786341761 (sc : alk. paper)

Subjects: LCSH: Atmospheric chemistry--Textbooks. | Atmospheric physics--Textbooks.

Classification: LCC QC879.6 .R58 2016 | DDC 551.51/1--dc23

LC record available at https://lccn.loc.gov/2016035655

British Library Cataloguing-in-Publication Data

A catalogue record for this book is available from the British Library.

Desk Editors: Suraj Kumar/Mary Simpson

Typeset by Stallion Press

Email: enquiries@stallionpress.com

Preface

This book, and its partner *Astrochemistry: From the Big Bang to the Present Day*, arose from a third year option course in Physical Chemistry entitled *Fundamentals of Astrochemistry and Atmospheric Chemistry* taught at the University of Oxford by myself and my colleague, Claire Vallance. *Astrochemistry: From the Big Bang to the Present Day* aims to provide an accessible introduction to this rapidly growing area of chemistry. The book relates the chemical history of the Universe, our Solar System, and our planet, looks in some detail at the 'alien' chemistry occurring in interstellar gas clouds, and also details the experimental and theoretical laboratory-based methods that have allowed us to gain deeper insight into the chemistry occurring in these regions. *Atmospheric Chemistry: From the Surface to the Stratosphere* considers in detail the physics and chemistry of our contemporary planet, and in particular its atmosphere, explaining the chemistry and physics of the air that we breathe, that gives rise to our weather systems and climate, soaks up our pollutants and protects us from solar UV radiation.

About the Author

Grant Ritchie is an Associate Professor of Physical and Theoretical Chemistry in the Department of Chemistry at the University of Oxford. Grant holds B.A and D.Phil degrees from the University of Oxford, working in the area of chemical reaction dynamics. Previously he has held a Ramsay Memorial Research Fellowship and a Royal Society University Research Fellowship before being appointed to a University Lectureship in Physical and Theoretical Chemistry at the University of Oxford in 2006. Grant's research focuses upon the development and applications of trace gas detection, principally for atmospheric monitoring and breath analysis; optical trapping, manipulation and spectroscopy of colloidal systems; and chemical reaction dynamics. Grant has published over 110 scientific papers in internationally recognized peer reviewed journals. Grant is also the John and Patricia Danby Fellow in Chemistry at Worcester College, Oxford, and responsible for undergraduate tutorial teaching across the whole of the Oxford physical chemistry course. Grant has previously co-authored the textbook *Physics for Chemists*, a title within the Oxford Chemistry Primers series.

Acknowledgements

I would like to thank Katherine Manfred and Michele Gianella for constructing multiple versions of many of the figures shown in this book and for answering all of my (very) daft questions about LaTeX with more courtesy than they deserved — your patience is truly appreciated. I would also like to thank Kim Whittaker whose enthusiasm for atmospheric chemistry was a large driver in convincing me to write this text, and Gus Hancock and Rob Peverall for their proofreading and helpful suggestions. I would also like to thank my research group who have put up with me as I have loudly worked my way through various versions of the text — you deserve *super scooby bonus points*!

I would also like to thank Claire Vallance, for her unstinting patience as I (repeatedly) dragged my heels on this project and for sharing a common vision for both our lecture course and this project. Laurent Chaminade, Catharina Weijman, and Mary Simpson from World Scientific also deserve considerable gratitude for their unending patience as agreed deadlines came and went — I take sole responsibility! Both Claire and I would also like to thank the undergraduate students at the University of Oxford who have attended our lecture course, asked questions that made us think, and provided feedback that enabled us to improve both our lecture courses and these books. We hope that they have enjoyed the lectures as much as we have!

Finally, I would like to thank Linzi, Ethan, and Erin who have provided me with the love, support, and freedom to pursue my academic interests at often the most strange and unsociable hours. I truly could not have asked for more.

Contents

List of Figures

List of Tables

Chapter 1

The Physical and Chemical Properties of the Earth's Atmosphere

Introduction

The Earth's atmosphere behaves as a dynamic fluid which can support a variety of motions on length scales spanning a few metres to the circumference of the entire planet. The atmosphere is also dynamic in the sense that its chemistry and physics are influenced not only by the photochemistry and physics induced by solar radiation, but also by interactions with the biosphere (all living organisms), the lithosphere (the crust and upper mantle), the hydrosphere (all water), and the cryosphere (a sub-division of the hydrosphere concerning only ice). Such interactions have important consequences, not least for the Earth's climate.

99.9% of the present-day atmosphere comprises N_2, O_2, the noble gases (mainly Ar), and H_2O. All of these species apart from water have a constant abundance, with water present in variable amounts depending upon the atmospheric conditions, latitude and altitude. The remaining 0.1% of the atmosphere consists of *trace gases*, including CO_2, CH_4, and others. Table 1.1 lists the abundances (expressed as mole fractions) and atmospheric lifetimes of the most important trace gases.[1] While the

---

[1]There are a number of ways of reporting atmospheric abundances. For example, the abundance of a species A may be reported as a mixing ratio, mr_A, which is simply the mole fraction of the species. Typically, as in Table 1.1, the mole fraction is reported as a *volume mixing ratio*, namely the fraction V_A/V of the total volume V taken up by species A, with units of parts per million by volume (ppmv) or parts per billion by volume (10^9) (ppbv). For an ideal gas these are equivalent, and $mr_A = n_A/n = V_A/V = p_A/p$, where n_A, V_A, p_A are the number of moles, the (partial) volume and partial pressure of species A in a sample of n moles of air of volume V and pressure p. The mixing ratio and

Table 1.1   The abundances and lifetimes of
the major constituents of the atmosphere.

Constituent	Mole fraction	Lifetime (yr)
N_2	0.781	1.6×10^7
O_2	0.209	9×10^3
Ar	9.3×10^{-3}	4.5×10^9
CO_2	400 ppmv	5
H_2O	$(0-4) \times 10^{-2}$	5 days
CH_4	1.8 ppmv	10
H_2	550 ppbv	4
N_2O	320 ppbv	150
CO	40–200 ppbv	0.2
O_3	20–80 ppbv	0.05
C_2H_6	1 ppbv	0.2
SO_2	100 pptv	5 days
NO_2	100 pptv	2 days

concentrations of trace species are many orders of magnitude smaller than those of N_2 and O_2, such species are present in chemically significant quantities, and have an important influence on many atmospheric processes. For example, the OH radical is the most important oxidant in the lower regions of the atmosphere, despite only being present at a concentration of *ca.* $10^6 \, cm^{-3}$ (or mixing ratio of 10^{-13} at standard temperature and pressure, STP).[2] In addition to their chemical role, trace gases can also affect the physical properties of the atmosphere. The most important example is provided by CO_2, a species that has a large effect on Earth's radiation budget despite its modest abundance of ~400 ppmv. The chemical and physical properties of OH and CO_2 will be the subject of detailed discussion in Sections 4.2 and 2.3, respectively.

Trace gases have both natural (e.g. biogenic, geologic, and oceanographic) and anthropogenic (e.g. fossil fuel burning and agriculture) sources, with the result that they may not be distributed evenly throughout the atmosphere. The physical and chemical lifetimes of each trace species compared with the timescale of mass transport within the atmosphere are important in determining the degree of spatial inhomogeneity. Such considerations determine how far industrial emissions are likely to spread

---

volume mixing ratio should be distinguished from the *mass mixing ratio*, μ_A, defined as $\mu_A = m_A n_A / m$ where m_A is the molecular mass of species A and m is the total molecular mass of the air sample.

[2]STP is defined as a pressure of 1 atm, temperature of 273 K, and number density of *ca.* $2.7 \times 10^{19} \, cm^{-3}$.

from their source. The factors that determine the lifetimes of species in the atmosphere are considered later in this chapter in Section 1.6. Before then, we will consider the physical structure of the atmosphere in terms of the variation in atmospheric pressure and temperature with altitude. We will investigate the role of moisture in the atmosphere, the factors governing atmospheric stability and transport, and then turn our attention to the kinetics pertinent to physical and chemical processing in the atmosphere. In Chapters 3 and 4, we will consider the chemistry of the two lowest regions of the atmosphere, the troposphere and the stratosphere, in more detail. While much of the chemistry of the atmosphere occurs in the gas phase, particulate matter in the form of aerosols are also an important atmospheric constituent with significant effects on both the chemical and physical properties of the atmosphere. Their effects will be considered in Chapter 5.

1.1 Structure of the Atmosphere

As shown in Figure 1.1, the atmospheric temperature does not simply decrease monotonically with increasing altitude above the surface, but instead displays a much more complex behaviour. The temperature profile is used to categorise various different regions within the atmosphere in terms of a series of layers. The detailed origins of the temperature profile will be considered in Section 1.3.

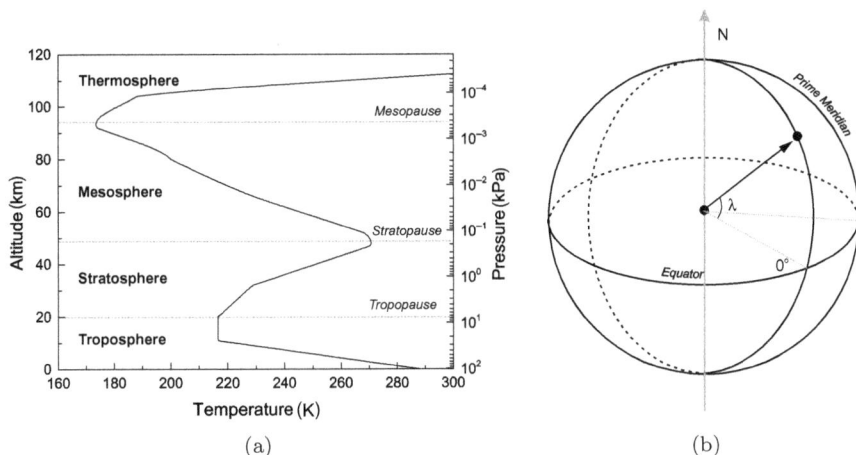

Figure 1.1 (a) The temperature and pressure structure of the Earth's atmosphere and (b) the definition of latitude.

1.1.1 The Troposphere

The bulk (90%) of the atmospheric mass lies in the troposphere, spanning the altitude range from 0 to 15 km. Within the troposphere, the temperature generally decreases with increasing altitude. In this region, the atmosphere is warmed by absorption of radiation from the Earth's surface, an effect that diminishes with increasing altitude. The temperature profile within the troposphere results in convection currents that make the troposphere relatively turbulent, and its contents well mixed. For the same reason, the troposphere is the region of the atmosphere that is responsible for weather. As we shall see, the longer term climate, and climate change, are determined primarily by radiant processes near the surface, and are also directly influenced by biogenic and anthropogenic emissions into the troposphere.

The sublayer of the troposphere closest to the Earth's surface, at altitudes below around 1 km, is known as the *planetary boundary layer* (PBL). The PBL is the region within which most pollutants are emitted into the atmosphere, from a variety of sources. Pollutants can become trapped within this layer by local weather conditions, with the result that high levels of pollution can build-up on timescales of a day or less, decreasing air quality to the point where there are direct negative effects on animal and plant life. We consider such effects in Section 4.7.

Fortunately, the majority of chemical pollutants are destroyed within the troposphere, principally by reaction with OH (day) and NO_3 (night) radicals. Such oxidation chemistry prevents most gases emitted at the surface of the Earth from accumulating to levels that are toxic or potentially damaging to others regions of the atmosphere such as the ozone layer. This ability to oxidise/chemically process emissions is an expression of the troposphere's *oxidising capacity*, and will be considered in more detail in Section 4.2.

At any given altitude, the temperature varies with latitude, λ (as defined in Figure 1.1(b)), and within the troposphere the *zonally averaged*[3] temperature generally decreases with increasing latitude. As can be seen in Figure 1.2, the *meridional* temperature gradient is considerably larger in the winter hemisphere, where the polar region is in darkness. The highest

---

[3] In this context, *zonal* has the meaning "along a circle of latitude" or "in the west–east direction", while *meridional* has the meaning "along a meridian" or "in the north–south direction".

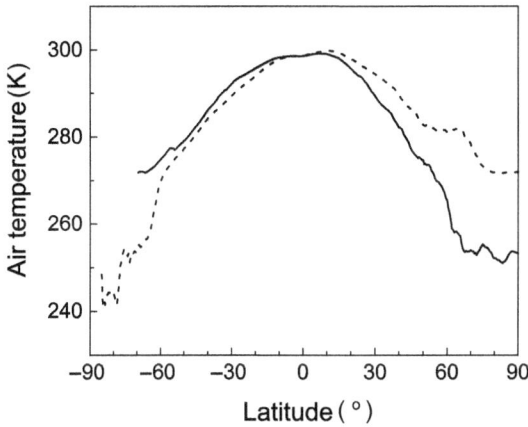

Figure 1.2 Zonally averaged surface temperature as a function of latitude. The two data sets are monthly averages for June (dotted line) and December (solid line) 2013.

Note: Analyses and visualisations used in this figure were produced with the Giovanni online data system, developed and maintained by the NASA GES DISC.

altitude (and hence lowest temperature) within the troposphere defines the *tropopause*. Consist with Figure 1.2, the altitude at which the tropopause occurs decreases markedly with latitude, ranging from *ca.* 15 km at the equator ($\lambda = 0°$), where the temperature is *ca.* 200 K, to 8 km at the poles ($\lambda = \pm 90°$) where the temperature is *ca.* 210–220 K.

As noted earlier, air in the troposphere is well mixed through convection. Above the tropopause there is a temperature inversion, marking the beginning of the stratosphere, and convection is no longer effective. The tropopause is therefore a relatively impermeable barrier to vertical transport, and material from the troposphere mixes slowly, if at all, with the stratosphere. As we shall see later in our discussion of stratospheric chemistry, the air in the stratosphere is extremely dry, indicating that material that does pass between the troposphere and the stratosphere enters through the coldest regions of the tropopause ($\sim$200 K) near the tropics. These regions act as a tropical "cold trap", freezing out any moisture in the air passing through to higher altitudes.

1.1.2 The Stratosphere

The stratosphere spans the altitude range from around 10 or 15 km up to $\sim$50 km, and is defined by its inverted temperature profile; temperature increases with altitude. Again, the temperature also varies with latitude, being warmest over the summer pole and coldest over the winter pole. The inverted temperature profile is a result of strong absorption of UV radiation by a layer of ozone molecules, O_3.[4] The detailed chemistry of the ozone layer will be considered further in Chapter 3. Convection is inhibited in an inverted temperature gradient, with the result that the stratosphere is relatively stable, with long vertical transport times. This property of the stratosphere has the unfortunate consequence that once pollution has reached the stratosphere, it stays there for a long time. Furthermore, as the vertical transport of matter is suppressed in the stratosphere, the physical behaviour of this region is dominated by radiative processes.

1.1.3 The Upper Layers of the Atmosphere

The top of the stratosphere is termed the *stratopause*, and above this region the temperature again decreases with altitude. The region directly above the stratosphere is called the *mesosphere*. In this region, the much lower molecular number densities means that there is no ozone to speak of (and no heating mechanism by ozone UV absorption) and hence both convective motion and radiative processes are important.

The *mesopause* defines the boundary between the mesosphere and another region of inverted temperature gradient, called the *thermosphere*. In this region, there is considerable dissociation and ionisation of O_2 and N_2 by short-wavelength solar and cosmic radiation, yielding a plasma of atoms, molecules, free electrons, and ions. The charged particles present in the thermosphere interact with the Earth's magnetic field, and temperatures in the thermosphere can change markedly in response to variations in solar activity. The mesosphere and thermosphere both contain ionised atomic and molecular species, and are often grouped together into a region extending from around 60 to 1000 km in altitude termed the *ionosphere*.

The Sun's upper atmosphere, or corona, produces a constant outward-flowing stream of high-energy charged particles, UV and X-ray radiation. The Earth's magnetic field deflects most of the charged particles before

---

[4] As discussed in *Astrochemistry: From the Big Bang to the Present Day* by C. Vallance, the development of the ozone layer and its ability to absorb UV radiation was critical to the evolution of life on Earth.

they reach Earth, but the high-energy radiation causes significant ionisation of the upper reaches of the Earth's atmosphere. While only half of the ionosphere (i.e. the region experiencing daytime) is irradiated by radiation from the Sun at any time, there are still sufficient cosmic rays to cause significant ionisation within the ionosphere during the night, though at a lower level than during the daytime.[5] The level of charge within the ionosphere is directly related to the intensity of incoming solar and cosmic radiation, and has practical importance due to its significant effects on communications technology, influencing radio propagation both between distant sites on Earth, and between satellites and Earth.

The ionosphere can be separated into three distinct regions according to the local ionisation behaviour. These regions are shown in Figure 1.3, and for historical reasons are named the D, E, and F regions, in order of increasing altitude. The extent of ionisation within a given region represents the balance between the effects of: (i) the type and intensity of ionising radiation reaching the region; (ii) the rate of electron–ion recombination, which depends on the local atmospheric pressure/number density; and (iii) the local atomic and molecular composition.

Within the D layer, some 60–90 km above the surface of the Earth, ionisation of NO by Lyman-α radiation ($\lambda = 121.5$ nm) occurs (N_2 and O_2 ionisation may also occur in periods of high solar activity) but the rate of

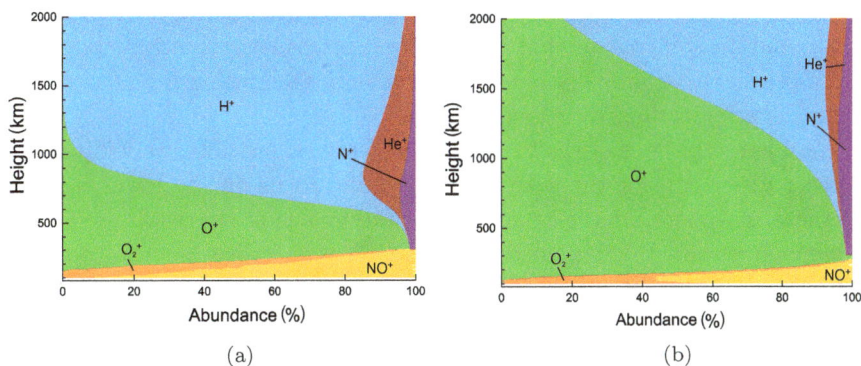

Figure 1.3   The composition of the ionosphere during the (a) night and (b) day.

---

[5] See *Astrochemistry: From the Big Bang to the Present Day* by C. Vallance for further discussion of cosmic rays.

recombination is high, and so net ionisation is low, with typical ion number densities of *ca.* 10^3–10^4 cm^{-3}. At night, this region is not ionised at all. By comparison, ionisation of O_2 dominates in the E layer (90–150 km), and the charge density is some one to two orders of magnitude higher than in the D region, though again significantly less at night. The F layer extends from an altitude of around 150 km to more than 500 km and is the most highly charged region of the ionosphere. The ion chemistry in this region is dominated by the formation of O^+, although at higher altitudes the number density of oxygen ions decreases and lighter ions such as hydrogen and helium become dominant. During the daytime, the F layer can sometimes be subdivided further into two layers labelled F_1 and F_2, with the F_2 layer remaining during both day and night.

The effect of the various layers within the ionosphere on radio communications is determined by their relative charge densities. The charge density in the D layer is normally not sufficient to interact significantly with radio waves, and signals are for the most part simply transmitted (with some losses) through to the E layer. Within the E and F layers, the charge density is high enough that propagating radio waves force free electrons into oscillation at the same frequency as the radio wave. If the electrons re-radiate the energy then it is transmitted back to Earth as a refracted wave. However, if electron–ion recombination occurs on a faster timescale than re-radiation (this commonly occurs within the D layer, leading to day-time signal losses) then the radio signal is damped and transmission efficiency is reduced. The high-charge density within the F layer relative to the E and D layers means that in practice it is this layer that is responsible for most skywave propagation of radio waves, facilitating high-frequency radio communications over long distances.

While the physics of the ionosphere is truly remarkable, we will curtail further discussion for the sake of brevity, and focus on the lower reaches of the atmosphere for the remainder of the following chapters. As a final comment, we note that the chemistry of the ionosphere has much in common with the chemistry of interstellar gas clouds as described in the first volume of this series,[6] and the interested reader is directed to a number of excellent reviews of the chemistry of this region.[7]

---

[6] *Astrochemistry: From the Big Bang to the Present Day* by C. Vallance.

[7] See for example A.M. Pavlov, *Surv. Geophys.* **33** 1133 (2012), J.M.C. Plane *et al.*, *Chem. Rev.* **115** 4497 (2015), and the text book *Ionospheres: Physics, Plasma Physics, and Chemistry* 2nd edition by R. Schunk and A. Nagy (Cambridge, 2009).

1.2 Atmospheric Pressure: The Hydrostatic Equation, Mixing Ratios, and Moisture

1.2.1 The Hydrostatic Equation

Atmospheric pressure, p, falls with increasing altitude, z. It is relatively straightforward to derive the *hydrostatic equation*, which quantifies this phenomenon, and which is based on balancing the opposing forces on a region of the atmosphere arising from the pressure gradient and gravity:

$$\mathrm{d}p = -\rho(z)g\mathrm{d}z \qquad (1.1)$$

where $\rho(z)$ is the air density at altitude z, and g is the acceleration due to gravity. The hydrostatic equation has the solution

$$p(z) = p_0 \exp(-z/H) \qquad (1.2)$$

where p_0 is the pressure at $z = 0$ (i.e. at the Earth's surface), and $H = RT/Mg$ is *scale height*, with R the gas constant, T the temperature, and M the average molar mass of the air $(28.9\,\mathrm{g\,mol^{-1}})$. The derivation and solution of the hydrostatic equation are outlined in Appendix A.

The scale height as defined is dependent upon mass, and it might therefore be supposed that each chemical species in the atmosphere will have a different scale height. However, air is uniformly mixed up to altitudes of 90 km, and there is no separation of heavy/light molecules into layers by mass. This homogenisation is ensured by rapid turbulent vertical mixing of air in the troposphere, discussed further in Section 1.4. Only at altitudes above 100 km does significant gravitational separation of gases take place, with lighter gases becoming enriched at higher altitudes. At the extremes, very light molecules such as H_2 can even leave the atmosphere. However, gravitational separation affects only a tiny fraction of the total mass of the atmosphere: the hydrostatic equation predicts that 99.9% of the atmospheric mass is found at altitudes below 50 km.

Equation (1.2) implies that the pressure drops by a factor of e over the scale height H. For our atmosphere, H takes values between 6 and 7.5 km, depending on the local temperature, with an average value of $\sim 7\,\mathrm{km}$. The hydrostatic equation can also be reformulated in terms of number density, n, rather than pressure, yielding the solution $n(z) = n_0 \exp(-z/H)$. Note that this expression can be applied to well-mixed gases such as N_2, O_2, CO_2, and CH_4, but not to gases such as ozone, whose variation with altitude depends on other factors. The altitude dependences of pressure and number density are shown in Figure 1.4.

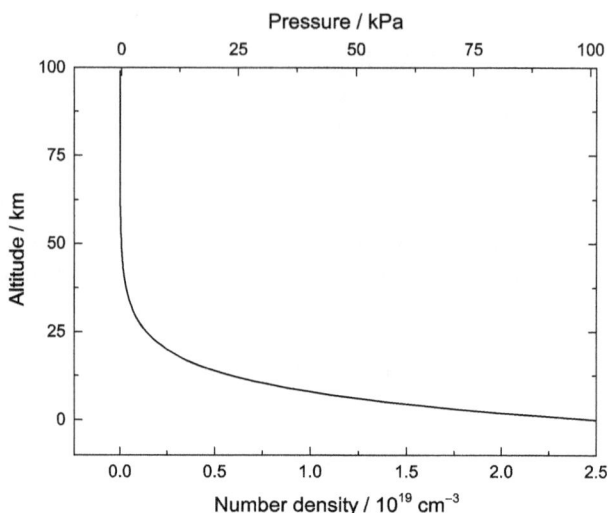

Figure 1.4   Altitude dependence of pressure and number density for an isothermal atmosphere.

The exponential reduction in both number density and pressure with altitude has the consequence that the rates of chemical reactions occurring in the atmosphere will slow down drastically with increasing altitude. The rate constants for the fastest (diffusion-controlled) neutral–neutral bimolecular processes are of the order of $2 \times 10^{-10} \, \text{cm}^3 \, \text{s}^{-1}$, while rate constants for neutral termolecular processes,[8] such as $O + O_2 + M \rightarrow O_3 + M$ (with M a third body, usually N_2 or O_2) are typically in the range 10^{-34}–$10^{-31} \, \text{cm}^6 \, \text{s}^{-1}$. At the lower end of this range, termolecular reactions thus occur at *ca.* 10^{-5} times the bimolecular collision rate at ground level, but some 10^{-8} times the bimolecular collision rate at an altitude of 50 km, where the bimolecular collision rate is itself *ca.* 10^{-6} times lower than at ground level.

Note that the discussion so far has assumed that the atmosphere exists in a state of hydrostatic balance, with a perfect balance between the upward

---

[8] Except at the very highest pressures, nominally termolecular reactions actually occur in two steps. In the reaction $O + O_2 + M \rightarrow O_3 + M$, the O and O_2 react in the first step to form highly vibrationally-excited ozone. This can either dissociate back to reactants, or undergo collision with a "third body". Transfer of vibrational energy to the third body in the collision yields stable O_3 products. The kinetics of termolecular reactions will be discussed further in Section 1.6.3.

force resulting from the pressure gradient and the downward force due to gravity. The atmosphere is very close to the condition of hydrostatic balance for most of the time. However, the balance can sometimes be lost in localised regions, leading to localised instabilities. As we will learn in Section 1.4, this leads to the generation of convection currents, which act to restore a stable atmospheric configuration.

1.2.2 Mixing Ratios and Column Abundances

It is clear from the preceding discussion that the absolute concentration of any species will change with altitude, regardless of whether any chemistry or transport occurs. However, the mixing ratio, *mr*, defined in the introduction to this chapter, *does not* have any intrinsic dependence on altitude. Any change in the mixing ratio of a species can therefore be attributed to changes in chemistry or transport. As an example, Figure 1.5 shows a stylised mixing ratio of methane as a function of altitude. In the troposphere, methane is well mixed and is relatively inert, having a lifetime of 10 years, and its mixing ratio is therefore constant throughout the troposphere. Within the stratosphere, however, CH_4 undergoes increased photochemical loss and slower rates of vertical transport. CH_4 is not well mixed in this region and its mixing ratio therefore decreases with altitude as shown.

The hydrostatic equation for number density can be used to calculate the *column amount* or *column density* of a particular species j, U_j. This is

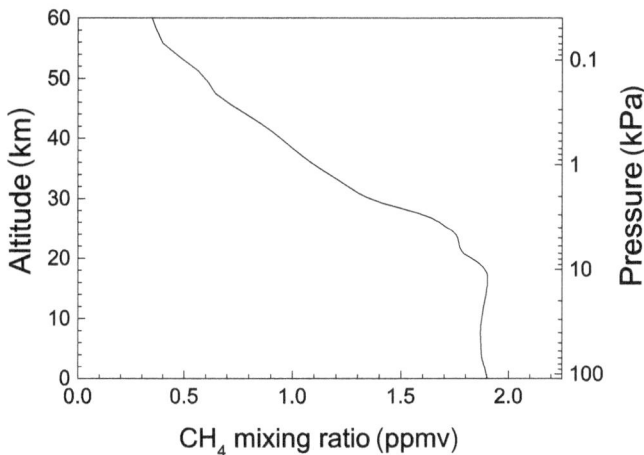

Figure 1.5   The mixing ratio of methane as a function of altitude.

determined by integrating the number density from the Earth's surface to
the top of the atmosphere (TOA).

$$U_j = \int_0^\infty n(z)\mathrm{d}z = Hn_0 \qquad (1.3)$$

The column amount has units of molecules cm^{-2}. Similarly, the *overhead
column amount* $U_{j,x}$ at an altitude x can be defined as

$$U_{j,x} = \int_x^\infty n(z)\mathrm{d}z = Hn_x \qquad (1.4)$$

Column densities can be calculated for any atmospheric species, whether
well mixed or not. In the case of ozone, which is not well mixed, the column
amount is conventionally defined in terms of a special unit called the *Dobson
unit* (DU), equivalent to $2.69 \times 10^{16}\,cm^{-2}$. One DU is the thickness, in 10
micron units, that the total ozone column would occupy at a temperature
of $273\,K$ and pressure of $1\,atm$ (very different conditions from those found
in the region of the ozone layer); typical column densities for ozone are *ca.*
$300\,DU$, corresponding to an ozone column only $3\,mm$ thick under these
(unrealistic) conditions.

1.2.3 Water in the Atmosphere

Although water is only a minor component of the atmosphere, its presence
in low but variable amounts has profound consequences for atmospheric
structure and dynamics. Atmospheric water affects atmospheric dynamics
through a variety of mechanisms. For example, as we will see in more
detail later, the adiabatic cooling of a "moist" air parcel can result in the
temperature of the parcel falling sufficiently far such that condensation
occurs. The heat released during this phase change gives the wet air parcel
extra buoyancy to ascend to a higher altitude, with the net result that
moisture in the atmosphere enhances convection in the troposphere. The
presence of water is also key to the Earth's weather systems, with phase
changes leading to precipitation in the form of rain, hail, and snow. Over
longer timescales, H_2O acts as a natural greenhouse gas, leading to warming
of the Earth's surface, with the formation of clouds further affecting the
radiative properties of the atmosphere. Finally, the formation of ice particles
in the atmosphere can facilitate important surface reactions which affect
for example, ozone distributions over Antarctica. We now consider the

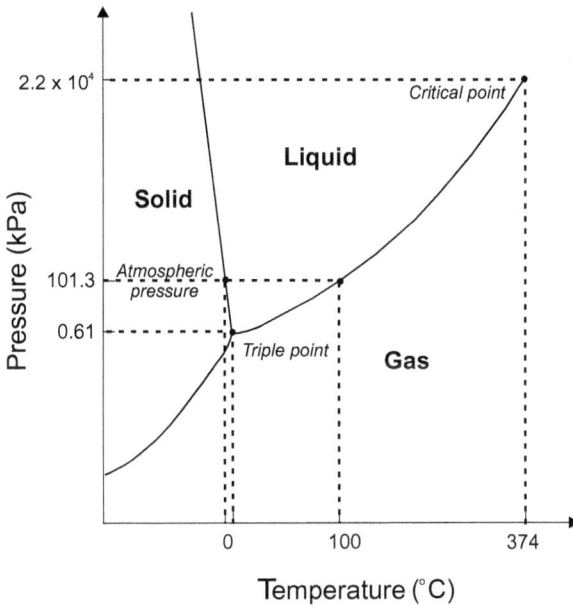

Figure 1.6   The phase diagram of water. [Note: not to scale]

thermodynamic properties of water that are of importance in atmospheric chemistry and physics.

A stylised phase diagram for water is shown in Figure 1.6. There are large regions of temperature–pressure space within which only a single phase is present, and certain combinations of (p, T) for which two phases are in equilibrium; the latter define the phase boundaries where melting/freezing, boiling/condensation, and sublimation occur. Elementary thermodynamics tells us that the slopes of the phase boundaries, $\mathrm{d}p/\mathrm{d}T$, are determined by the entropy change, ΔS, associated with the phase change, and the corresponding volume change, ΔV, via the *Clapeyron equation*

$$\frac{\mathrm{d}p}{\mathrm{d}T} = \frac{\Delta S}{\Delta V} = \frac{\Delta H}{T\Delta V} \tag{1.5}$$

where ΔH is the change in enthalpy associated with the phase transition, and T is the transition temperature.

In most cases, we are interested in the liquid–gas phase transition. For this case, we can make the simplification that the volume of the liquid phase is negligible relative to that of the gas, and therefore that to a very good approximation, ΔV is simply equal to the volume of the gas. Making the further approximation that the gas behaves ideally (obeying the ideal gas law, $pV = nRT$), we arrive at the *Clausius–Clapeyron* equation

$$\frac{\mathrm{d}\ln p}{\mathrm{d}T} = \frac{\Delta H_{\mathrm{vap}}}{RT^2} \tag{1.6}$$

where ΔH_{vap} is the enthalpy of vapourisation. This equation allows the vapour pressure of water (or any other substance) to be determined as a function of temperature. The equilibrium vapour pressure at a given temperature is known as the *saturated vapour pressure*.

While the Clausius–Clapeyron equation was derived above for water in isolation, the same arguments can also be applied to water vapour in the presence of air. In this case, the total pressure, p, is replaced by the partial pressure of water vapour, traditionally denoted e. The volume mixing ratio mr_w of water (assuming it behaves ideally) is simply e/p, and the phase transition is described by the equation

$$\frac{\mathrm{d}\ln e_{\mathrm{s}}}{\mathrm{d}T} = \frac{\Delta H_{\mathrm{vap}}}{RT^2} \tag{1.7}$$

where e_{s} is the *saturation vapour pressure*, i.e. the value of the partial pressure of water vapour above which condensation will occur. Integration of the Clausius–Clapeyron equation, assuming that ΔH_{vap} is independent of T, gives

$$e_{\mathrm{s}}(T) = A \exp\left(-\frac{\Delta H_{\mathrm{vap}}}{RT} + B\right) \tag{1.8}$$

where A and B are constants. Figure 1.6 clearly shows that e_{s} increases markedly with increasing temperature. At the Earth's surface, e is of the order of 10 mbar at a temperature of 283 K. As the atmospheric temperature reduces with altitude, then so does the moisture content.

Rather than reporting the absolute vapour pressure of water in air, e, it is common instead to report the vapour pressure relative to the saturation vapour pressure, e_{s}; this is the *relative humidity*, RH $= e/e_{\mathrm{s}}(T)$. Alternatively, the water content of air can be recast in terms of its *mass mixing ratio*, μ, defined as $\mu = \epsilon(mr_w)$ where $\epsilon = m_w/m = 0.622$ is the ratio of the molar mass of water vapour, m_w, to the mean molar mass, m, of moist air. While e remains less than $e_{\mathrm{s}}(T)$ saturation does not occur, no condensation or evaporation takes place, and μ is constant. However, if an air parcel rises sufficiently high in the atmosphere, its temperature and pressure can reduce to the point at which e becomes equal to $e_{\mathrm{s}}(T)$, and saturation occurs. The pressure at which this happens is called the *lifting condensation level* and the corresponding *saturation mixing ratio*, is

$$\mu_s(T, p) = \frac{e_s(T)\epsilon}{p} \tag{1.9}$$

with units of grams (of water) per kilogram (of air), $g\,kg^{-1}$. Typical values of μ_s range from 1 to $100\,g\,kg^{-1}$ depending upon both temperature and pressure. As will be discussed in Section 5.3.1, once the saturation point has been reached, this does not guarantee formation of liquid water. Instead, the vapour may become *supersaturated*. Under these conditions, it is usually necessary for small condensation nuclei to be present before liquid drops appear.

While we have focused our attention so far on the liquid–vapour phase transition, there are also solid–liquid and solid–vapour transitions. In the latter case the phase boundary is determined by the Clausius–Clapeyron equation but with ΔH_{vap} replaced by the enthalpy change for sublimation, ΔH_{sub}. In the former case the volume of each phase is similar and the phase boundary is determined by the Clapeyron equation,

$$\frac{dp}{dT} = \frac{\Delta S_{fus}}{\Delta V_{fus}} \tag{1.10}$$

where ΔS_{fus} and ΔV_{fus} are the entropy change and volume change associated with fusion/melting. The density of liquid water is greater than that of ice, with the consequence that the ice–liquid water phase boundary has a negative slope. Finally, we note that there exists a single pressure and temperature at which all three phases co-exist — this is known as the *triple point*, and occurs at a temperature of 273.16 K and a pressure of 611.7 Pa.

1.3 Atmospheric Temperature Profiles

1.3.1 The Tropospheric Temperature Profile: The Adiabatic Lapse Rate

As noted previously in Section 1.1, the troposphere is defined by a temperature that reduces with increasing altitude. This temperature gradient can be rationalised by considering the expansion work done by an air parcel as it rises through the atmosphere. For our purposes, we define an air parcel as a hypothetical volume of air to which may be assigned all of the basic dynamic and thermodynamic properties of atmospheric air. Clearly, the

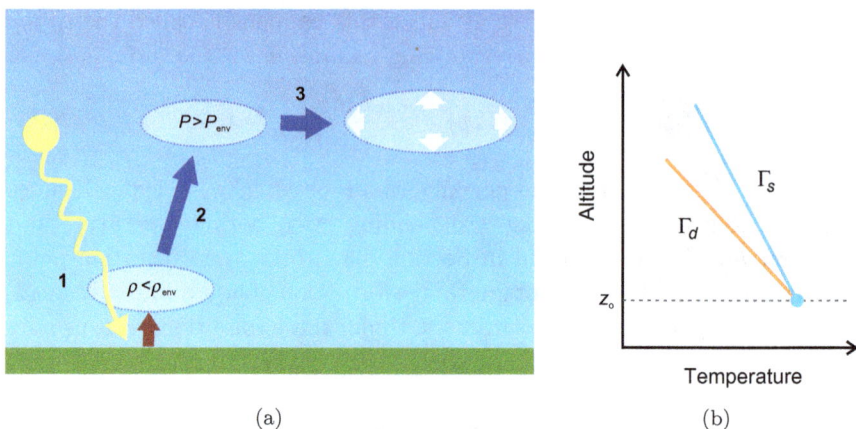

(a)   (b)

Figure 1.7   (a) Schematic of an air parcel rising: (1) solar radiation heats the surface which then re-radiates IR to heat the air parcel; (2) density of the parcel becomes lower than surrounding environment and parcel rises; (3) pressure of surrounding air is less than the parcel and parcel expands. (b) Comparison of dry (Γ_d) and saturated (Γ_s) adiabatic lapse rates.

parcel must be large enough to contain a very large number of molecules, but small enough so that its properties are approximately uniform. We also assume that the parcel is insulated from its surroundings, but that the parcel boundary is flexible, permitting the air inside to expand or contract as conditions dictate, and ensuring that the pressures outside and inside the parcel will always be the same.

Direct sunlight does not heat the atmosphere efficiently, but instead heats the surface of the Earth. Heating of the lower atmosphere occurs primarily through absorption of infrared (IR) radiation re-radiated from the surface, as illustrated in Figure 1.7. An air parcel undergoing heating near the surface becomes less dense than its surroundings, and so will rise to higher altitudes. As it rises into a region of lower pressure, it expands. To expand it must do work on the external environment, losing energy and cooling. Qualitatively, this simple model explains the reduction in temperature with increasing altitude in the troposphere.

Quantitatively, the rate of change of temperature with altitude, dT/dz, known as the *lapse rate*, can be calculated using elementary thermodynamics. The First Law of Thermodynamics states that the change in the internal energy of the parcel, dU, is given by

$$dU = dq + dw \tag{1.11}$$

where dq is the heat supplied *to* the air parcel and dw is the work done *on* the air parcel. Assuming that the process is adiabatic (i.e. that there is no transfer of heat into or out of the parcel) then d$q = 0$ and so

$$dU = dw = -pdV \qquad (1.12)$$

where p is the external pressure and dV is the change in volume of the air parcel as it expands.

Enthalpy is defined as $H = U + pV$, so d$H = dU + pdV + Vdp$. This can be combined with Equation (1.12) to give d$H = Vdp$. Using the hydrostatic equation and the definition of density, we then have

$$dH = -V\rho g dz = -Mg dz \qquad (1.13)$$

where M is the molar mass of the air. Differentiating with respect to temperature at constant pressure yields

$$\left(\frac{dH}{dT}\right)_p = -Mg\left(\frac{dz}{dT}\right)_p \qquad (1.14)$$

Recognising the heat capacity at constant pressure, $C_p = (dH/dT)_p$, the above expression can be rearranged to give the *dry adiabatic lapse rate*

$$\Gamma_d = -\left(\frac{dT}{dz}\right)_p = \frac{Mg}{C_p} \approx 9.7 \text{ K km}^{-1} \qquad (1.15)$$

The dry adiabatic lapse rate often overestimates the true or *environmental lapse rate*, Γ_{env}, in the real atmosphere. Typically $\Gamma_{env} \approx 6$ K km^{-1}. The discrepancy between Γ_d and Γ_{env} arises because real air contains water vapour. Moist air has a larger heat capacity than dry air, leading to a reduced lapse rate. However, Equation (1.15) for the lapse rate still holds as long as the correct heat capacity for the air parcel is employed. More importantly, as an air parcel rises to regions of lower temperature, the saturation vapour pressure will gradually reduce, and the RH will increase. If the parcel has a high moisture content, then the RH may exceed 100%, and water will condense in the air parcel. Condensation is an exothermic process, and the enthalpy of vapourisation is released into the air parcel, causing warming. Under these *saturated* conditions, the rising air parcel experiences both expansion cooling and condensation warming with the net result that the parcel still cools on ascent, but at a slower rate. This new rate is called the *saturated adiabatic lapse rate*, Γ_s. A mathematical expression for Γ_s is derived in Appendix B. Importantly, Γ_s is always less than Γ_d, and can be highly variable. If the air is cold, then Γ_s is only slightly less than Γ_d, but Γ_s can fall as low as $0.35\Gamma_d$ at high temperatures, such as those found near the ground in tropical humid conditions.

The strong dependence of the lapse rate on the moisture content of a given air parcel has the result that the lapse rate varies considerably with spatial location, time and altitude. Under some circumstances, Γ_{env} can even be negative. For example, at night, very close to the ground, the air temperature often increases with altitude over a small range. This temperature inversion provides stability within the PBL, and can cause air parcels containing pollutants to be trapped close to the surface. We will return to this point in Section 4.7.

1.3.2 The Potential Temperature

As an air parcel is transferred adiabatically to a new altitude, its temperature, pressure and density all change. In order to predict its subsequent behaviour, it is important to know whether the parcel is more or less dense than the new environment in which it finds itself (note that "warmer" does not necessarily mean "lighter" or "less dense"). Regions of equivalent density can be identified by evaluating a property known as the *potential temperature*, θ, defined as the temperature that a given air parcel would reach if it was moved adiabatically to an altitude corresponding to some predefined constant pressure, p_0, usually taken to be 1 bar. Importantly, the potential temperature of a given air parcel *does not* change as it rises or falls through the atmosphere.

To derive an expression for θ, we recognise that for an adiabatic transfer, no heat is transferred into or out of the air parcel, and so $dq = 0$. The First Law of Thermodynamics then yields

$$\mathrm{d}H = C_p\mathrm{d}T = V\mathrm{d}p = \frac{RT}{p}\mathrm{d}p \tag{1.16}$$

and therefore

$$C_p\mathrm{d}T = \frac{RT}{p}\mathrm{d}p \tag{1.17}$$

where we have assumed that the gas behaves ideally. Integrating this expression between the limits $(T(z), p(z))$ and (θ, p_0), corresponding to the current and final altitude of the air parcel, yields *Poisson's equation* for the

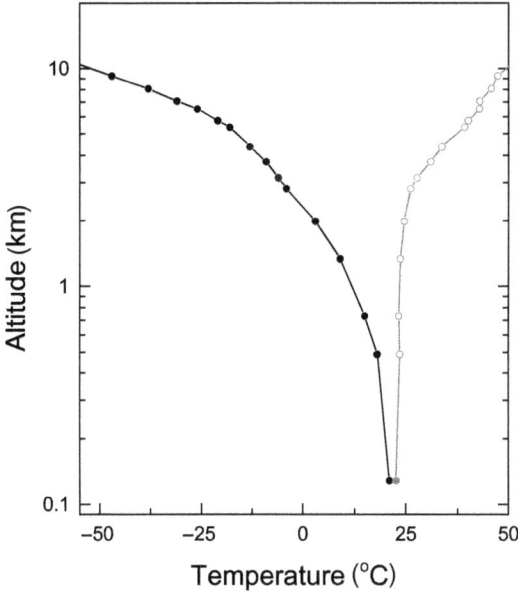

Figure 1.8   A plot of actual (filled circles) and potential (open circles) temperature as a function of altitude.

Note: The data pertain to measurements at Rochester, NY, on 3 May 2015.

potential temperature

$$\theta(z) = T(z) \left(\frac{p_0}{p(z)} \right)^{R/C_p} \tag{1.18}$$

with $R/C_p = 2/7$ for an ideal diatomic gas at typical atmospheric temperatures.

Regions of identical potential temperature have the same density, with an increase in θ corresponding to a decrease in density, and vice versa. As we shall see later in Section 1.5, the resulting θ *surfaces* (surfaces of constant θ) tend to define the paths along which air travels, with important implications for transport within the troposphere. An example of the potential temperature of the atmosphere as a function of altitude, z, is shown in Figure 1.8.

1.3.3 The Stratospheric Temperature Profile

The lapse rate in the stratosphere is negative, i.e. the temperature increases with altitude. The temperature inversion is explained by strong absorption of solar UV radiation by O_3 and O_2, which leads to their photolysis and subsequent exothermic reactions of the photoproducts. For example, UV absorption by ozone initiates the photolysis process

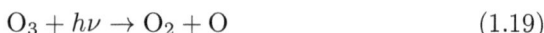

$$O_3 + h\nu \rightarrow O_2 + O \tag{1.19}$$

The atomic oxygen then goes on to react with O_2

$$O + O_2 \leftrightarrow O_3^*$$
$$O_3^* + M \rightarrow O_3 + M \tag{1.20}$$

where the species M is a "third body" (usually N_2 or O_2) capable of collisionally stabilising the highly internally excited nascent ozone, O_3^*, formed in the recombination of O and O_2. The overall result of these reactions is to convert UV radiation into thermal energy (i.e. into translational and/or internal motion of M and the newly formed O_3). A similar heating effect occurs following O_2 photolysis, but is less efficient. The rate of these processes is determined both by the number density of O_3 and O_2 and by the intensity of UV light. The former decreases with altitude, while the latter increases with altitude (since at lower altitudes much of the short-wavelength UV radiation has already been absorbed by ozone and oxygen at higher altitudes). The net result is that the overall rate of these processes, and therefore the atmospheric temperature, increases

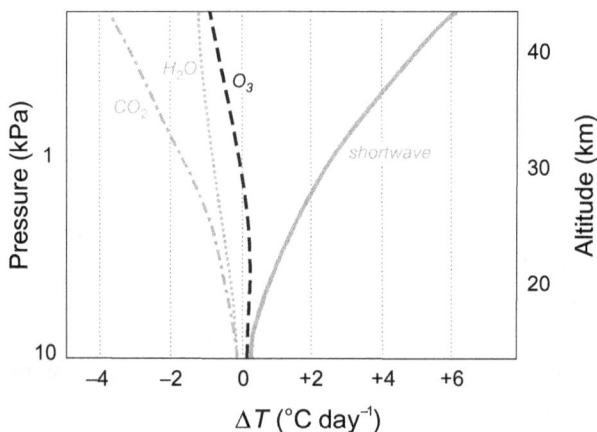

Figure 1.9   Radiative balance in the upper troposphere/stratosphere showing equilibrium radiative change due to shortwave absorption by H_2O, O_3, and CO_2 (solid) and longwave emission by H_2O (dotted), O_3 (dashed), and CO_2 (dash dotted).

Source: Adapted from S. Manabe and R.F. Strickler, *J. Atmos. Sci.* **21** 361 (1964).

with increasing altitude throughout the stratosphere, reaching a maximum temperature at the stratopause. As will be discussed in the next chapter, the stratosphere is very close to existing in radiative equilibrium, with a net radiative heating rate of zero. Radiative heating due to the absorption of solar radiation is almost exactly balanced by cooling via emission of long-wavelength radiation to space by H_2O, CO_2, and O_3 as illustrated in Figure 1.9. It should be noted that CO_2 makes the most important contribution to stratospheric cooling. We also note that, in comparison to the troposphere, there is a much smaller contribution to the radiation balance arising from radiation exchange between different atmospheric layers, and to a good approximation, this energy transfer mechanism can be ignored within the stratosphere.

1.4 The Vertical Stability of the Atmosphere: Buoyancy and Turbulence

1.4.1 Buoyancy

Within the troposphere, both the temperature and the pressure decrease with increasing altitude. However, the density, $\rho = p/RT$, may decrease or increase with altitude depending on the detailed form of the atmospheric temperature profile, $T(z)$. This has important consequences for the vertical stability of the atmosphere. When the density of the atmosphere decreases with increasing altitude, the system is stable. In this case, if a packet of air was to rise through the atmosphere, it would have a greater density than the surrounding air (negative buoyancy), and gravity would act to reverse the rise. However, occasionally localised regions will develop in which the density increases with increasing altitude. A rising air packet now has a lower density than the air it displaces, and will continue to accelerate in the vertical direction, experiencing *positive buoyancy*. The balance between the upward pressure gradient force and the downward gravitational force acting on the air packet is no longer maintained, and hydrostatic equilibrium has been lost. The upward acceleration, a, can then be calculated to be

$$a = -\frac{1}{\rho}\frac{\mathrm{d}p}{\mathrm{d}z} - g \tag{1.21}$$

The atmosphere is very close to the condition of hydrostatic balance for most of the time, except at isolated locations when the vertical profile becomes unstable as described above. Via the mechanism outlined above,

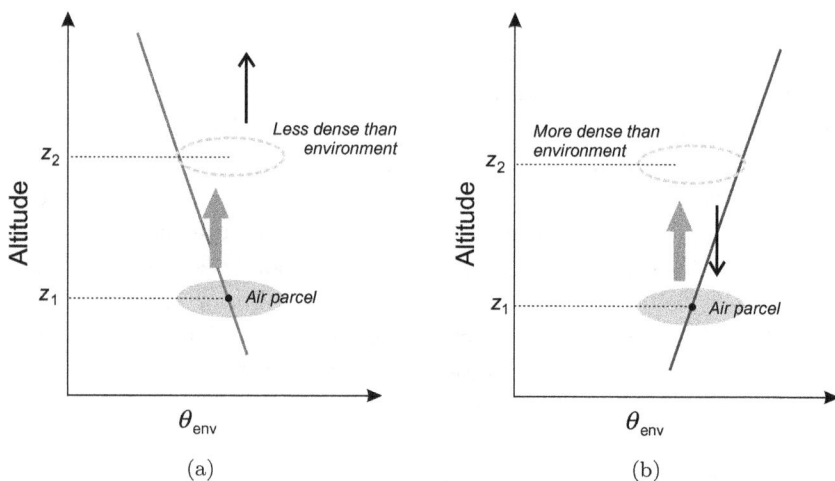

Figure 1.10 A schematic of how the potential temperature of the atmosphere dictates the vertical stability of an air parcel. In (a) $d\theta_{env}/dz < 0$ and the parcel is unstable, while in (b) $d\theta_{env}/dz > 0$ and the parcel is stable. The black arrows represent the resultant motion of the air parcel.

a region of unstable air will eventually evolve into a region of stable air, with the lowest densities at the highest altitudes. This process is known as *convection*, and occurs on a relatively short timescale of a few hours.

Stable and unstable air is often discussed quantitatively in the context of the potential temperature introduced in Section 1.3.2. Consider the case shown in Figure 1.10(a), in which the potential temperature of the environment, $\theta_{env}(z)$, decreases with increasing altitude, i.e. $d\theta_{env}/dz < 0$. Initially, the air parcel has the same potential temperature as its environment, $\theta(z_1) = \theta_{env}(z_1)$ (note that unsubscripted properties refer to the air packet, and properties with the subscript "env" refer to the surrounding environment). When the parcel is raised to altitude z_2, it undergoes adiabatic expansion such that its pressure drops to that of the environment at the new altitude, i.e. $p(z_2) = p_{env}(z_2)$. For an adiabatic process, θ is conserved and so

$$\theta(z_2) = \theta(z_1) = \theta_{env}(z_1) > \theta_{env}(z_2) \qquad (1.22)$$

(since $\theta_{env}(z)$ reduces on increasing altitude from z_1 to z_2).

Since $\theta(z) \propto T(z)$, we can conclude that $T(z_2) > T_{env}(z_2)$, and given that $p(z_2) = p_{env}(z_2)$, it must be true that $\rho(z_2) < \rho_{env}(z_2)$. The air parcel is less dense than its environment, and will continue to accelerate upwards through the atmosphere. The situation in which $d\theta_{env}/dz < 0$ is therefore statically unstable. Conversely, as shown in Figure 1.10(b), for the case in which $d\theta_{env}/dz > 0$, an air parcel that rises from altitude z_1 to z_2 will become more dense than its environment, and will therefore fall back to its original altitude. The air is then said to be statically stable. Similar arguments apply when an air parcel is lowered in altitude rather than raised. In the unstable case it will continue to accelerate downwards, whereas in the stable case it will tend to return to its starting altitude.

Instead of formulating ideas of stability in terms of the potential temperature, because we are assuming that the pressure within the air parcel is always equal to the pressure of the surrounding environment, it generally suffices simply to consider the temperature lapse rates, Γ and Γ_{env}, of the air parcel and its environment. If $\Gamma_{env} > \Gamma$ (i.e. the environmental temperature decreases more rapidly than the rate of adiabatic cooling as the air parcel rises), then the parcel will become warmer, and therefore less dense, than its environment, and the ascent will continue to greater altitude. This is an unstable condition. An alternative criterion for static stability is therefore that $\Gamma_{env} < \Gamma$. Both situations are illustrated in Figure 1.11(a)

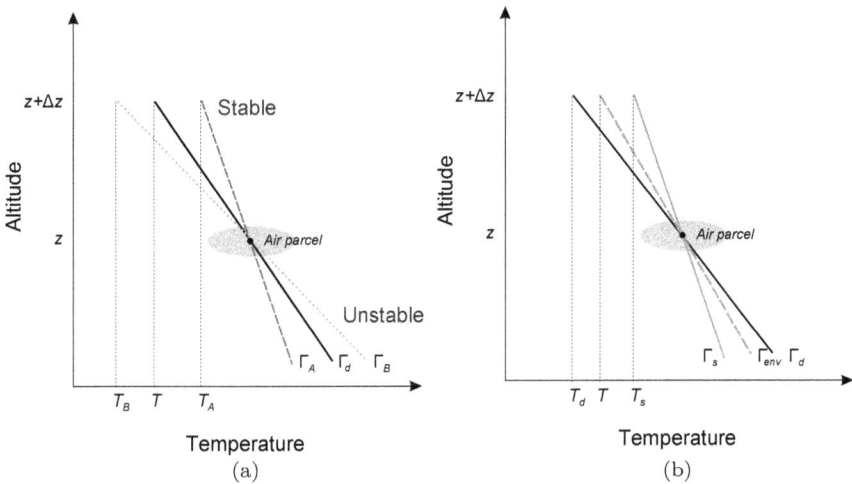

Figure 1.11   (a) Comparison of the dry adiabatic lapse rate of the air parcel, Γ_d, with the environmental lapse rate, Γ_{env}, for (A) stable (Γ_A, dashed line) and (B) unstable (Γ_B, dotted line) conditions and (b) the case of conditional instability in which $\Gamma_s < \Gamma_{env} < \Gamma_d$.

where Γ is taken to as the dry adiabatic lapse rate, Γ_d. It also follows that if $\Gamma_{env} < \Gamma_s$, then the air is stable even if the air is saturated. Conversely if $\Gamma_{env} > \Gamma_s$ then a saturated parcel is unstable. Interestingly, if $\Gamma_s < \Gamma_{env} < \Gamma_d$, a saturated parcel is unstable but an unsaturated one is stable; this situation is called *conditional instability* and is depicted in Figure 1.11(b).

The temperature gradient determines the vertical stability of the atmosphere and therefore the speed of vertical mixing. In the stratosphere, the positive temperature gradient with increasing altitude makes the stratosphere extremely stable with respect to vertical displacements. Vertical mixing in the stratosphere is therefore very slow, occurring on a timescale of months to years. In contrast, the negative temperature gradient within the troposphere makes it unstable with respect to vertical mixing. Air parcels that are displaced up or down are not subject to large restoring forces, and vertical mixing is fast, occurring on a timescale of hours to days, and leading to a variety of winds and weather. The troposphere in fact earned its name from this behaviour: the Greek word *tropos* means "turning", while *stratos* translates as "layered". In summary, the stable stratosphere acts as a lid on top of the unstable troposphere with the tropopause acting as a (leaky) barrier to transport.

1.4.2 Turbulence

Buoyancy in an unstable atmosphere accelerates air parcels both upwards and downwards, with *no* preferred direction of motion. This behaviour is modified by turbulence, providing the mechanism by which pollutants and other chemical species are dispersed. Consider a fluid (liquid or gas) flowing slowly through a circular capillary, as shown in Figure 1.12. Inside the capillary, the flow velocity is a function of the radial distance from the centre of the capillary, and the flow can be readily treated using standard hydrodynamics equations. This type of flow is known as *laminar flow*. If the

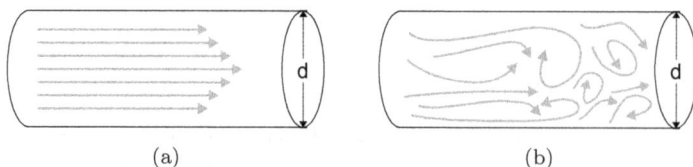

Figure 1.12   (a) Laminar and (b) turbulent flow through a tube of diameter d.

diameter of the capillary, d, or the flow velocity, v, is increased, a sudden change takes place at a particular critical diameter or flow velocity to a very different, chaotic state of motion called *turbulent flow*. Reynolds showed that flow becomes turbulent when a dimensionless parameter now known as the Reynolds number, Re, surpasses a certain critical value. The *Reynolds number* for a fluid of density ρ is given by

$$\text{Re} = \frac{\rho dv}{\mu} \tag{1.23}$$

where μ is the (dynamic) viscosity of the fluid. Laminar flow occurs for $\text{Re} \leq 2300$, while the regime $2300 \leq \text{Re} \leq 4000$ is described as transient.[9]

Flow within the atmosphere can be laminar or turbulent, depending on the atmospheric conditions. For example, on a calm evening, a small fire can create a well-defined column of smoke that can be recognised over a reasonable distance (a laminar plume), while by day the smoke disperses in random directions (turbulent flow) within (even) relatively light breezes. The most important example of turbulence within the atmosphere is the turbulence generated within the first kilometre or so from the Earth's surface by convection and wind (discussed further in the next section). However, it should be emphasised that the troposphere as a whole is subject to turbulence all the way up to the tropopause.

To develop a more quantitative model of turbulence, we write the flow velocity in a turbulent flow as the sum of the average flow velocity, $\mathbf{v}_0$, and fluctuating turbulent velocity components (u, v, w) in the (x, y, z) directions, respectively. For example, when the flow is in the x-direction, the instantaneous flow velocity is $(v_0 + u, v, w)$.[10] Because turbulence is chaotic, it is impossible to pre-specify (u, v, w) as a function of time. However, we can develop a statistical (rather than deterministic) description of the phenomenon. Turbulent fluctuations in the flow velocity occur at random and therefore the time-average of each of the turbulent components is zero, i.e. $\langle u \rangle = \langle v \rangle = \langle w \rangle = 0$. However, their *variances* are non-zero.[11] The average values of the products of any two turbulent velocity components, i.e. $\langle uv \rangle$, $\langle vw \rangle$, etc., are also in general non-zero. This implies that motion in

---

[9] For a review of the history of the Reynolds number see N. Rott, *Ann. Rev. Fluid Mech.* **22** 2 (1990).

[10] By convention, when describing flow in the atmosphere, the z-axis is defined as lying in the "upwards" direction (normal to the Earth's surface), x lies along the average flow (wind) direction, and y is therefore directed across the average flow direction.

[11] The variance for the x component of the flow velocity, for example, is equal to $\langle u^2 \rangle - \langle u \rangle^2$.

two directions is *correlated*, and reflects the existence of *eddies* of rotating and swirling fluid. This type of motion has important consequences for atmospheric transport.

It is common to quantify the rate of turbulent mixing by analogy with molecular diffusion. Fick's First Law of Diffusion states that the flux,[12] F, of material down a concentration gradient, $\partial C/\partial x$, is given by $F = -D\partial C/\partial x$, where D is the diffusion coefficient. Fick's Second Law of Diffusion allows calculation of the mean-squared displacement of a molecule, $\langle x^2 \rangle$, undergoing random motion as a function of time; for motion in one dimension this relationship is $\langle x^2 \rangle = 2Dt$.[13] Under atmospheric conditions the diffusion coefficient takes a value of *ca.* $0.1\,\mathrm{cm^2\ s^{-1}}$, and diffusion is therefore a very slow process on the macroscopic length scale of the atmosphere. Despite this conclusion, turbulence can still be described as a type of diffusion: whenever there is an imbalance in concentration, momentum, or temperature, the atmosphere responds to reduce such gradients, and does so via turbulence. By analogy with the molecular diffusion coefficient, D, we can define an *Eddy diffusion coefficient*, K, and for turbulent motion in the z-direction this yields

$$\langle z^2 \rangle = 2Kt \tag{1.24}$$

Within the PBL, K has a value of around $10^5\,\mathrm{cm^2\ s^{-1}}$ over land and $10^3\,\mathrm{cm^2}$ $\mathrm{s^{-1}}$ over the sea. The value of K also varies with the time of day, becoming larger in the morning and smaller at night.[14] With a K value of $10^5\,\mathrm{cm^2}$ $\mathrm{s^{-1}}$, it will take on average *ca.* 2 months for air to mix vertically over an altitude range of $10\,\mathrm{km}$. Species with lifetimes of this order and longer are therefore well mixed in the troposphere, whereas those with shorter lifetime display large vertical gradients in their abundance. Mixing within the PBL occurs within a period of a day or so, although we note that very rapid mixing in the first $100\,\mathrm{m}$ or so can occur on the order of an hour or less. In comparison, the exchange of air between the troposphere and the

---

[12]Flux is defined as the quantity of material passing through an imaginary surface of unit area per unit time.

[13]For the three-dimensional case $\langle \mathbf{x}^2 \rangle = 6Dt$.

[14]A familiar example of the difference between molecular diffusion and turbulent mixing can be found in the addition of cream to coffee. If the cream is poured very slowly, it will float on the coffee surface and remain separate. Eventually the cream will diffuse into the coffee to make a uniform mixture, but only on a timescale of the order of a month. Alternatively, the contents of the cup may be stirred, thereby inducing turbulent mixing on the timescale of a few seconds.

stratosphere is far slower (due to the temperature inversion) and occurs on a timescale of 5–10 years.

1.5 Horizontal Transport: Winds and Circulation

Circulation of the atmosphere transports heat, momentum, clouds, water, ozone, and pollution around the globe. The general circulation depends both on differential heating of the Earth's surface, and therefore the atmosphere, and on the vertical temperature profile of the atmosphere. The total incoming solar flux or *insolation* varies little with season, but as shown in Figure 1.13, there is a marked variation in solar flux reaching the surface of the Earth as a function of latitude, with the solar flux at the equator being higher than at the poles. In contrast, the outgoing longer-wavelength radiation escaping into space from the Earth's surface shows a much less extreme variation with latitude, a point we will return to in the next chapter. While energy conservation requires that the globally-averaged incoming absorbed solar radiation is balanced by the outgoing terrestrial radiation, the spatial differences in the incoming and outgoing intensity distributions result in a negative temperature gradient on moving from low to high latitudes.

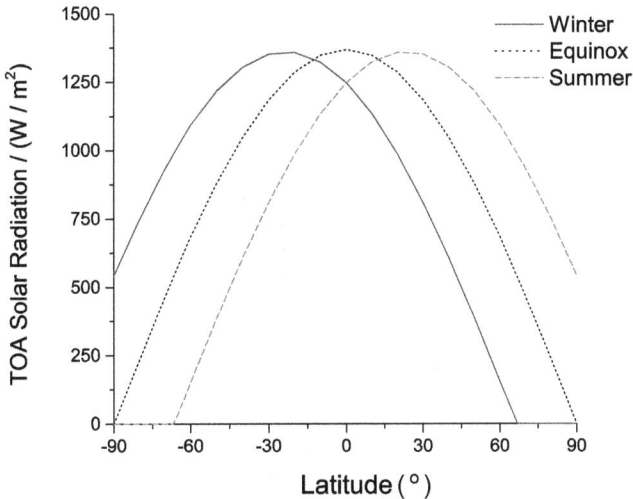

Figure 1.13 The variation of the solar flux at the TOA (modelled) as a function of latitude for different seasons. The maximum value of the distributions have been scaled to be 1360 W m^{-2}.

The horizontal temperature gradients caused by differential heating of the Earth's surface result in horizontal pressure gradients, which in turn leads to air flow from regions of high pressure (at or near the equator) to low pressure (towards the poles), commonly known as "wind". In addition to the force exerted by the pressure gradient, air flowing in the horizontal direction (i.e. parallel to the Earth's surface) also experiences a second force as a result of the *Coriolis effect*.[15] This tends to deflect a moving air parcel to the right of its direction of motion in the Northern hemisphere, and to the left in the Southern hemisphere.

As shown in Figure 1.14, when the Coriolis force and the pressure gradient forces balance, the air travels approximately parallel to isobars (lines of

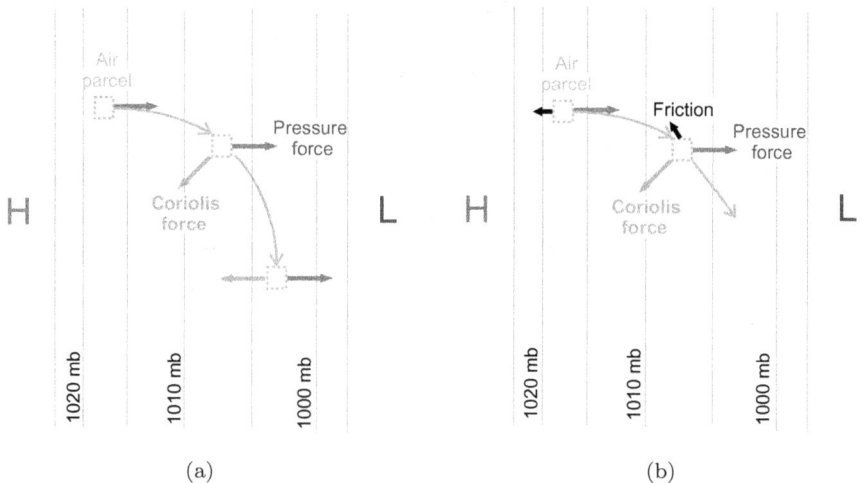

(a)   (b)

Figure 1.14   (a) The development of geostrophic flow and (b) ageostrophic flow. In both cases the air parcel initially moves in the direction of the pressure gradient before being subject to the Coriolis force. The size of the Coriolis force is dependent upon the angular velocity of the Earth's rotation, the speed of the air parcel in the rotating frame of reference and the latitude.

---

[15]The Coriolis effect describes the fact that an object moving in a straight line in a non-rotating frame (for example, imagine viewing air flowing from the equator to the poles over the Earth's surface as the Earth spins beneath it) will appear to experience a deflection in a direction opposite to the direction of rotation when viewed in the rotating frame (i.e. by someone standing on the surface of the Earth). Though the origin of the deflection is sometimes referred to as a "Coriolis force", this is not really accurate, as there is no force acting in the non-rotating frame. For this reason, the Coriolis force is sometimes referred to as a *pseudo-force*. A detailed mathematical description of the Coriolis effect can be found in the many excellent textbooks on classical mechanics.

constant pressure). Such conditions are said to constitute *geostrophic flow*. In the Northern hemisphere this means that the high pressure region is to the right of the flow, i.e. air flows clockwise around a centre of high pressure and clockwise around a centre of low pressure. The former situation is generally referred to as an *anticyclone* or a *High*, while the latter is known as a *cyclone* or a *Low*. In reality, the moving air near the surface of the Earth also experiences a frictional force as it moves over the varied surface topography, which opposes the direction of motion. The result is that the Coriolis force does not quite balance the pressure gradient force, and the resulting *ageostrophic flow* is only nearly parallel to isobars. The surface friction decreases the flow velocity and thereby decreases the Coriolis force, meaning that air is deflected towards the region of low pressure. The combined result of these effects is that strong zonal winds with speeds of the order of $10\,\mathrm{m\ s^{-1}}$ are prevalent within the atmosphere; these winds can circumnavigate the globe on a timescale of a few weeks. By contrast, meridional transport is slower, with wind speeds an order of magnitude less, so that it takes *ca.* 2 months for air at mid-latitudes to exchange air with the tropics.

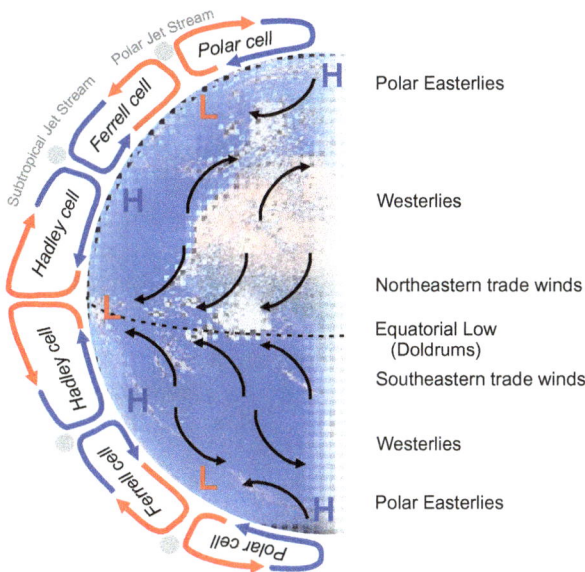

Figure 1.15   Hadley, Ferrell, and Polar circulation cells with wind patterns.

On a slowly rotating planet we would therefore expect warm air to rise at low latitudes, move polewards via the mechanisms outlined above, and descend there, forming a flow structure known as a *Hadley cell*. In fact, the Earth's rotation is sufficiently fast that the increasing size of the Coriolis force with latitude causes air flow in the high-altitude branches of the Hadley cell to break down into an unstable flow, leading to the development of three circulation cells in each hemisphere. These are known as the *Hadley*, *Ferrell*, and *Polar* cells, and are shown in Figure 1.15. Under geostrophic flow conditions, transport between circulation cells is difficult, as the air must lose angular momentum. In practice, this can be achieved through surface friction in ageostrophic flow conditions. The more varied surface topography in the Northern hemisphere, and correspondingly increased surface friction, makes this process more efficient than in the Southern hemisphere. The result is that poleward transport of air towards Antarctica is reduced relative to poleward transport to the Arctic, and is an important factor in promoting formation of the ozone hole. The ozone hole will be discussed in detail in Section 3.3.

Given that geostrophic flow is driven by the latitudinal heating gradient, it should be unsurprising that inter-hemispheric transport occurs relatively slowly, on a timescale of around a year. Finally, it should also be noted that large-scale circulation is responsible for less than half of the total polewards transport by the atmosphere. The majority is the result of turbulent transfer, waves and storms, collectively called *eddy diffusion*.

1.6 Chemical Kinetics in the Atmosphere

The Earth's atmosphere can be thought of as a giant photochemical reactor which can exchange material with the surface and also transport matter around the globe via large-scale motions. In the following discussion, we will consider the general processes that determine the atmospheric lifetimes of trace species. The lifetime of a species within the atmosphere relative to the timescale associated with transport determines how effectively it will be distributed to locations remote from its source.

1.6.1 Atmospheric Lifetimes

As shown in Table 1.1, trace gases vary widely in both their abundance and their lifetime. The lifetime, τ, of a substance deposited into the atmosphere (assuming no further deposition), is determined by the rates of both

Table 1.2   Physical and chemical sinks for a selection of atmospheric constituents.

Sink process	Minor constituents removed
Physical processes	
Dry deposition to water and land surfaces	SO_2, O_3, HNO_3, CO_2, H_2O_2
Wet deposition in precipitation	HCl, H_2SO_4, NH_3, SO_2, HNO_3, H_2O_2, aerosols
Chemical processes	
Oxidation by OH	VOCs, CO, SO_2, NO_2, H_2S, H_2O_2, DMS
Photolysis	O_3, NO_3, NO_2, H_2O_2, HCHO, CH_3I
Cloud and aerosol reactions	H_2SO_4, NH_3, SO_2, HNO_3

chemical and physical loss processes. Common examples of chemical loss processes, known as *sinks*, include photolysis, oxidation reactions (the main gas-phase reactions involving attack by OH, O_3, and NO_3), and cloud and aerosol reactions. Physical loss processes include wet and dry deposition, both of which will be discussed later in this section. A selection of sink processes and the atmospheric constituents that they remove are presented in Table 1.2.

The overall loss rate of species A follows a first-order rate law,

$$\frac{d[A]}{dt} = -k[A] \qquad (1.25)$$

where the overall rate constant k is the sum of the first-order (or pseudo-first-order)[16] rate constants for the various loss processes.

$$k = k_{\text{photolysis}} + k_{\text{reaction}} + k_{\text{deposition}} + \cdots \qquad (1.26)$$

The rate law may be integrated to give an expression for the concentration of A as a function of time:

$$[A] = [A]_0 e^{-kt} = [A]_0 e^{-t/\tau} \qquad (1.27)$$

where $[A]_0$ is the initial concentration of A. The lifetime is the time required for the concentration of A to drop to $1/e$ of its initial concentration, and is simply the inverse of the overall first-order rate constant for loss of A.

Gases with long lifetimes (stable molecules such as chlorofluorocarbons (CFCs), CH_4, N_2O) are uniformly distributed in the troposphere, and so are

---

[16] If species A is lost through reaction with another atmospheric gas, B, present at constant concentration then the overall rate law, $d[A]/dt = -k[A][B]$, may be written $d[A]/dt = -k'[A]$, where $k' = k[B]$ is a pseudo-first-order rate constant.

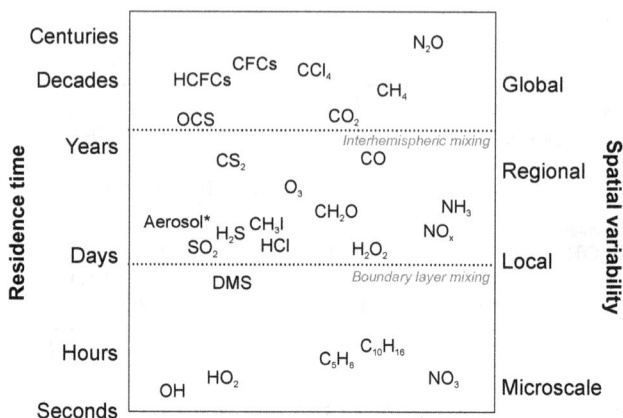

Figure 1.16   Residence time and spatial distribution of some important atmospheric species. CFCs, HCFCs and DMS denote chlorofluorocarbons, hydrochlorofluorocarbons and dimethyl sulphide, respectively.

affected by atmospheric dynamics. Conversely, species with short lifetimes, such as radicals, SO_2, NO_2, show temporal and spatial variations in their abundance, depending on the spatial and temporal distribution of their sources, and tend to be present in low concentrations. Natural and anthropogenic emissions may be highly localised (e.g. power plants, volcanoes) and consist of both reactive and inert species. Different atmospheric constituents can therefore display a wide range of spatial distributions and lifetimes as shown in Figure 1.16.

The atmospheric concentration of a given species depends on the rate at which it is introduced into the atmosphere and on its lifetime. Figure 1.17(a) shows a general kinetic scheme in which species A is introduced into the atmosphere at a constant rate E, and can be lost by both chemical and physical routes. The rate of change of [A] with time is now

$$\frac{d[A]}{dt} = E - k_c[A] - k_p[A]$$

$$= E - k_{tot}[A] \tag{1.28}$$

where k_{tot} is the sum of the rate constants k_c and k_p for chemical and physical removal, respectively. Imposing the initial conditions that at $t = 0$, [A] $= 0$, this first-order differential equation has the solution

$$[A](t) = \frac{E}{k_{tot}}(1 - e^{-k_{tot}t}) \tag{1.29}$$

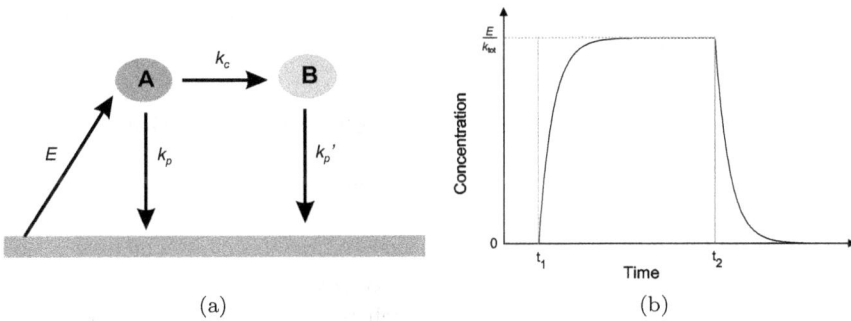

Figure 1.17   (a) Generalised kinetic scheme showing emission (E) of trace species **A** with loss rates to physical deposition (k_p) and chemical conversion to **B** (k_c), which can then also be lost by physical removal processes (k'_p). (b) Time dependence of [A] in the atmosphere where t_1 is the time that emission begins and t_2 is the time at which it ends.

As can be seen from Figure 1.17(b), the concentration of A builds up in the troposphere to a maximum value of E/k_{tot}. If emissions suddenly cease, so that $E = 0$, then [A] falls exponentially, with time constant $\tau = 1/k_{\text{tot}}$.

1.6.2 Physical Loss Processes

Dry deposition of a species refers to removal at the Earth's oceanic and terrestrial surfaces. The rate of dry deposition depends upon the rate of transport through the PBL (whose height varies between day and night) from the atmosphere to the Earth's surface and on the rate of reaction and/or adsorption at the surface. If surface removal is efficient, dry deposition will result in a gradient in the trace gas concentration with altitude, with lower concentrations near the surface where removal is occurring. The lifetime of a trace gas with respect to deposition is given by $\tau_{\text{dep}} = h/v_g$ where h is the height to which the trace gas is well mixed and v_g is the deposition velocity, i.e. the speed at which the species approaches the surface. The rate of deposition is therefore

$$-\frac{d[A]}{dt} = \frac{h}{v_g}[A] = [A]\tau_{\text{dep}} \tag{1.30}$$

For aerosols, v_g depends on particle size, with larger particles falling out more quickly. Wet deposition occurs through two separate mechanisms, *rain out* and *wash out*. During rain out the trace gas is removed by dissolution of water-soluble species into cloud droplets (and possible subsequent reaction within the droplet), which then grow and eventually fall to the surface as rain, snow, etc. Wash out differs in that the trace gas is simply taken up into the falling precipitation.

1.6.3 Bimolecular and Termolecular Chemical Reactions

Bimolecular reactions in the atmosphere may occur via either direct or indirect (complex-forming) mechanisms. The dynamics of both types of reaction, and the types of potential energy surface over which the two types of reaction occur, have been examined previously.[17] The thermal rate constants, $k(T)$, of direct reactions, which have no minimum in potential energy along the reaction coordinate, tend to be well described by the Arrhenius equation over the range of temperatures found in the atmosphere

$$k(T) = A \exp(-E_a/RT) \tag{1.31}$$

where A is the pre-exponential factor, and E_a is the activation energy. In contrast, the rate constant for an indirect reaction may show a complex dependence on temperature and pressure. Common examples of direct reactions in the atmosphere include H atom abstraction reactions of OH with saturated hydrocarbons. An example of an indirect reaction is the important tropospheric reaction $HO_2 + NO \rightarrow OH + NO_2$. The reaction is second-order, and displays a negative temperature dependence, implying the formation of an intermediate complex. A second example is the reaction $OH + HNO_3 \rightarrow H_2O + NO_3$. In this case, the rate coefficient is pressure-dependent, and again shows a negative temperature dependence, which is more marked at higher pressures. A reaction profile for this reaction is shown in Figure 1.18. The reduction in rate constant with increasing temperature results from the formation of an energised complex, which becomes less stable at higher temperature and therefore more likely to dissociate back into reactants at a given pressure than to collide with a stabilising third body, M.

The pressure dependence is attributable to the competition between complex-formation and decomposition. A complex-forming reaction between reactants A and B has the generic mechanism

$$A + B \underset{k_{-a}}{\overset{k_a}{\rightleftharpoons}} AB^* \tag{1.32}$$

$$AB^* + M \overset{k_q}{\longrightarrow} AB + M \tag{1.33}$$

---

[17]See Chapter 5 of *Astrochemistry: From the Big Bang to the Present Day.*

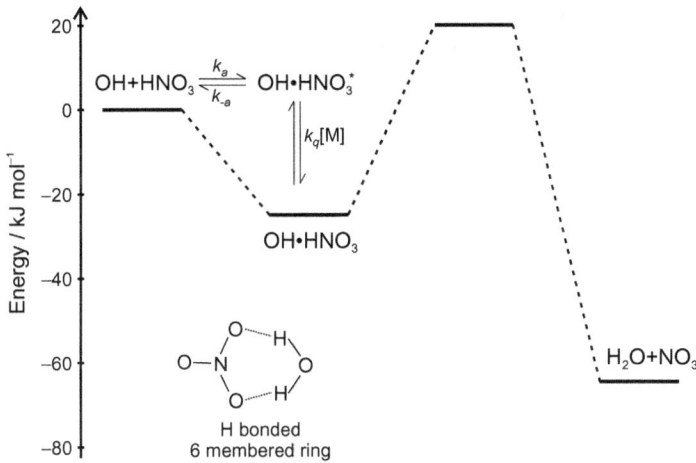

Figure 1.18 Reaction profile for OH + HNO$_3$.

Applying the steady-state approximation (SSA)[18] to the complex AB* yields the following rate law for formation of the stable AB product.

$$\frac{\mathrm{d}[A]}{\mathrm{d}t} = \frac{k_a k_q [M][A][B]}{k_{-a} + k_q [M]} = k_{\mathrm{eff}}[A][B] \qquad (1.34)$$

The effective rate constant, k_{eff}, is pressure dependent, due to its dependence on the concentration of the third body, M. In fact, even the reaction order is pressure dependent. At high pressure, such that $k_q[M] \gg k_{-a}$, the rate law simplifies to give a limiting value for the rate equal to $k_a[A][B]$, and the reaction becomes second-order. In this case, the first step in the mechanism is rate limiting, as the high concentration of M means the second step is rapid. At low pressures, when $k_q[M] \ll k_{-a}$, the rate is equal to $(k_a k_q / k_{-a})[M][A][B]$, and the reaction is third-order overall, with quenching of the excited complex to give the stable AB product now the rate-determining step in the mechanism. The pressure

---

[18]The SSA can be applied to an intermediate in a reaction mechanism when its rate of removal greatly exceeds its rate of formation. Under these conditions, the concentration of the intermediate rapidly reaches a constant value, which is maintained until the end of the reaction when the reactants have been used up. This situation applies to most reactive intermediates. However, if one or more of the reactants from which the intermediate is formed are very short-lived then the rate of production of the intermediate cannot be treated as constant, and the SSA breaks down.

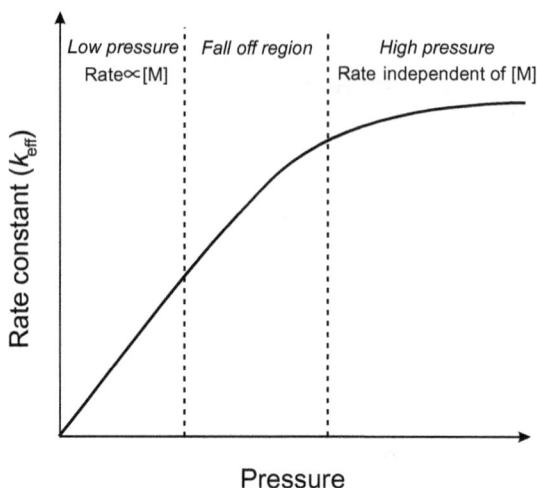

Figure 1.19   Pressure dependence of the rate constant for a generic complex-forming reaction involving a third body. Third-order behaviour (pressure dependent) occurs at low pressure and second-order behaviour (pressure independent) at high pressures.

dependence of k_{eff} is shown in Figure 1.19. The rate constants of many important atmospheric reactions, for example $O + O_2 + M \rightarrow O_3 + M$ and $OH + SO_2 + M \rightarrow HSO_3 + M$, exhibit similar behaviour.

As noted previously, any short-lived species, including reactive intermediates, do not need to be included in models of transport, as they will never travel far from their source before being removed by reaction. However, there are some pairs of species that exist in equilibrium, as a result of rapid interconversion between two or more species. In this case, the sum of the concentrations of the two species may remain fairly constant, even though the lifetimes of the individual species are very short. Such pairs, which include O and O_3 (referred to collectively as O_x, with $\tau(O)$ and $\tau(O_3) \ll \tau(O + O_3)$), and NO/NO_2 (or NO_x), are said to be in *photochemical steady state*. The relative amounts of each species in an (A,B) pair may be determined by considering the kinetics of the equilibrium between them,

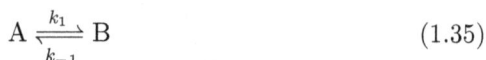

$$A \underset{k_{-1}}{\overset{k_1}{\rightleftharpoons}} B \tag{1.35}$$

with k_1 and k_{-1} the rate constants for the forward and reverse reaction. If only species A is present initially, at concentration $[A]_0$, then the rate of change of [B] is given by

$$\frac{d[B]}{dt} = k_1[A] - k_{-1}[B]$$

$$= k_1([A]_0 - [B]) - k_{-1}[B]$$

$$= k_1[A]_0 - (k_1 + k_{-1})[B] \qquad (1.36)$$

Making the substitution $y = k_1[A]_0 - (k_1 + k_{-1})[B]$ yields the following solution for the time-dependent concentration of B

$$[B](t) = \frac{k_1[A]_0}{k_1 + k_{-1}}(1 - e^{-(k_1+k_{-1})t}) \qquad (1.37)$$

We can find the steady-state value of [B] by letting t tend towards infinity, giving $[B] = k_1[A]_0/(k_1+k_{-1})$. The relative values of the forward and reverse rate constants therefore determine the steady-state composition, with the steady state being reached on a timescale $\tau = 1/(k_1 + k_{-1})$.

1.7 Summary

At this point, we have provided an overview of the physical structure of the atmosphere, the variation in atmospheric temperature and pressure with altitude and latitude, and the consequences of temperature and pressure gradients for chemical transport. We have also discussed the most important features pertaining to the rates of the physical and chemical processes of most prominence in the atmosphere. In the following chapters, we will study the chemistry of the troposphere and stratosphere in detail, highlighting the importance of both gas-phase and heterogeneous processes. Before that however, we consider in detail how the chemical structure of our atmosphere modifies the solar spectrum to which the Earth is continuously exposed, and show that the presence of trace gas and aerosol species have a significant impact upon the temperature of both the atmosphere and the planetary surface.

1.8 Questions

1.8.1 Essay-style Questions

Q1.1: Discuss with examples any termolecular reactions of importance in the Earth's atmosphere. Indicate how the rate constant for a termolecular reaction will change as a function of temperature and pressure.

Q1.2: The terrestrial temperature falls with increasing altitude within the troposphere up to *ca.* 12 km. Above this altitude the temperature rises in the stratosphere. In this region, the maximum ozone concentration occurs. Explain these observations.

Q1.3: Explain why the effects of convection in the Earth's atmosphere become small at high altitude.

Q1.4: Indicate qualitatively how an increase in tropospheric temperature might affect the composition of the troposphere.

Q1.5: Define *potential temperature* and explain its importance in considering the vertical stability of the atmosphere. Estimate the potential temperature, θ, at the tropopause assuming that the pressure there is 0.1 bar and the surface temperature is 290 K, stating any assumptions you make.

1.8.2 Problems

P1.1: Check that 1 DU corresponds to an ozone column of *ca.* 2.7×10^{16} molecules cm^{-2}.

P1.2: Consider the hypothetical situation in which the average temperature of the atmosphere is decreased by 1 K and the heat thereby released is given to the top 100 m of the ocean, whose average temperature rises by an amount ΔT everywhere. Find ΔT given that the specific heat capacity, c_p, of sea water is $4.2 \, kJ \, K^{-1} \, kg^{-1}$ and that of air is $1.04 \, kJ \, K^{-1} \, kg^{-1}$. Take the radius of the Earth to be 6400 km.

P1.3: (a) Calculate the air density at the summit of Mount Everest assuming that the pressure and temperature were $3.13 \times 10^4 \, Pa$ and 234.5 K respectively.

(b) A parcel of air at $8 \times 10^4 \, Pa$ with a temperature of 280 K is surrounded by air with a temperature of 278 K and is lifted to a pressure of $7 \times 10^4 \, Pa$ where the air temperature is 272 K. Is the parcel warmer or cooler than its environment when arriving at 7×10^4 Pa? What have you assumed in your calculation?

P1.4: (a) Find the density of moist air with a *specific humidity* (the ratio of the mass of water vapour to the total mass of the air sample) of $15\,\text{g kg}^{-1}$, at temperature $288\,\text{K}$ and pressure 1.01×10^5 Pa, and compare it with the density of dry air at the same temperature and pressure.

(b) The *virtual temperature*, T_v, is the temperature of air having the same density as a sample of moist air at the same pressure. Show that T_v is given by the expression $T_\text{v} = T \left(\frac{1+q_\text{v}/\epsilon}{1+q_\text{v}} \right) \approx$ $T(1 + 0.6q_\text{v})$ where the mass mixing ratio $q_\text{v} = M_\text{v}/M_\text{d}$, $\epsilon =$ m_v/m_d and T is the temperature of the moist air. M_d and M_v are the masses of dry air and water vapour respectively and m_d and m_v are the corresponding molecular masses.

(c) Calculate the virtual temperature of the air sample in part (a).

(d) An air parcel of moist air has a temperature of $297.9\,\text{K}$. It is neutrally buoyant sitting in an environment that has a temperature of $298\,\text{K}$ and a mixing ratio of $10\,\text{g kg}^{-1}$. Given that neutral buoyancy requires the air parcel and the environment to have the same virtual temperature, calculate the mixing ratio in the air parcel.

P1.5: (a) Prove that

$$\frac{C_\text{v}\mathrm{d}T}{T} + \frac{n R\mathrm{d}V}{V} = 0$$

for a reversible adiabatic expansion of n moles of an ideal gas.

(b) Hence show that if C_v is independent of temperature, $pV^\gamma = K$ where K is a constant, and $\gamma = C_p/C_v$.

(c) According to the kinetic theory of gases, the internal energy of an ideal monatomic gas is given by $U = 3/2nRT$. Obtain expressions for the molar isochoric (constant volume) and isobaric (constant pressure) heat capacities of the gas. Why, physically, should the latter be larger than the former?

(d) One mole of an ideal gas initially at $300\,\text{K}$ is expanded adiabatically and reversibly from 20 to 1 bar. What is the final temperature of the gas? What is the work done as a result of the expansion?

P1.6: (a) Starting from $dG = VdP - SdT$ derive both the Clapeyron equation and the integrated form of the Clausius–Clapeyron equation, which shows how the vapour pressure of a solid or liquid varies with temperature. State clearly any assumptions that you make.

(b) The barometric pressure at the top of Mount Everest is *ca.* 250 mmHg. The normal boiling point of water is 100°C and its enthalpy of vapourisation is 44.9 kJ mol^{-1}. Estimate the temperature at which water would boil at the top of the mountain.

(c) Estimate the vapour pressure of water at 120°C.

P1.7: According to a simple thermodynamic model, the temperature T of any planetary atmosphere should vary with altitude z according to:

$$-\frac{dT}{dz} = \frac{g}{c_p}$$

where g is the acceleration due to gravity and c_p is the constant pressure heat capacity per unit mass. Estimate c_p for the Earth's atmosphere below 80 km altitude, explaining the basis of your estimate and any necessary assumptions. Hence, calculate the dry adiabatic lapse rate. [Take g to be 9.8 m s^{-2}.]

P1.8: Derive the relationship given below for the altitude dependence of atmospheric pressure, $p(z)$, in terms of the pressure p_0 and temperature T_0 at sea level, assuming that the temperature decreases uniformly with altitude at a rate Γ K m^{-1} i.e. $T = T_0 - \Gamma z$, and that gravity is a constant:

$$\left(\frac{p(z)}{p_0}\right) = \left(\frac{T_o - \Gamma z}{T_0}\right)^{\frac{M_R g}{R\Gamma}}$$

R is the gas constant, M_R is the molar mass of the atmosphere and g is the acceleration due to gravity. You are reminded that the hydrostatic equation is $\frac{dp}{dz} = -\rho g$ where ρ is the density of the gas. The gas may be treated as ideal.

P1.9: (a) A parcel of moist air is lifted adiabatically from the Earth's surface. Derive its lapse rate for the unsaturated part of the ascent.

(b) At the lifting condensation level, saturation occurs. Show that $c_p dT + L d\mu + g dz = 0$ where μ is the liquid water mixing ratio, L is the latent heat of vapourisation per unit mass, z is altitude and c_p is the *specific* heat capacity of air at constant pressure, i.e. the heat capacity per unit mass. Hence, show that the lapse rate for the saturated ascent can be written as

$$\Gamma_s = -\frac{dT}{dz} = \left(\frac{g}{c_p}\right)\left(1 + \frac{L\mu_s}{R_a T}\right)\left(1 + \frac{L^2 \mu_s}{c_p R_v T^2}\right)^{-1}$$

where μ_s is the saturation mixing ratio, R_a is the specific gas constant for air, and R_v is the specific gas constant for water vapour. The specific gas constant, R_i is the gas constant per unit mass of air, and therefore depends upon the precise composition of the sample of air under consideration.

(c) At a temperature of $10°C$ and pressure of 10^4 Pa, μ_s is *ca.* 20 g kg^{-1} and the enthalpy of vapourisation per unit mass is 2260 kJ kg^{-1}. Estimate a value for Γ_s.

P1.10: (a) A parcel of air passes through a deep convective cloud and reaches the level of neutral buoyancy (LNB) in the upper troposphere. We assume that it is dry and that $\theta = 350$ K. Since the parcel arrives at the LNB with upward momentum it overshoots the LNB. The environmental profile is stable with a potential temperature (θ_{env}) lapse rate that is constant with altitude at $d\theta_{env}/dz = 5 \times 10^{-3}$ K m^{-1}, and so the parcel will return to the LNB and oscillate. The parcel's acceleration is related to the buoyancy as follows

$$d^2(\Delta z)/dt^2 = g\left(\frac{\theta - \theta_{env}}{\theta_{env}}\right)$$

where Δz is the vertical displacement from the LNB.
Show that $d^2(\Delta z)/dt = -N^2(\Delta z)$ where $N^2 = g(d\theta_{env}/dz)/\theta_{env}$.

(b) If $N^2 > 0$ then the solution of this second-order differential equation is $\Delta z(t) = A\sin(Nt) + B\cos(Nt)$; the vertical displacement of the parcel oscillates in time and the system is stable. In this case, N is known as the *Brunt–Väisälä frequency* (or buoyancy frequency) and the period of oscillation is $\tau = 2\pi/N$. Derive the period of oscillation of the parcel.

P1.11: HCFC-22 is a hydrofluorochlorocarbon whose concentration has been steadily building in the atmosphere at a rate of 8.5×10^7 kg per year.

(a) The average rate at which HCFC-22 is input to the atmosphere is 2.5×10^8 kg per year due to industrial emissions, significantly larger than the build-up rate. Determine its atmospheric lifetime, assuming that the current mass of HCFC-22 in the atmosphere is 2.2×10^9 kg.

(b) The only atmospheric loss of HCFC-22 is by reaction with OH, the rate constant for which is 4.7×10^{-15} cm^3 molecule^{-1}s^{-1}. Calculate the concentration of OH.

(c) If the rate constant of 4.7×10^{-15} cm^3 molecule^{-1}s^{-1} is appropriate at 298 K and 1 bar, how will this constant change as the reaction takes place at increasing altitudes within the troposphere?

(d) Calculate the expected timescale for vertical mixing of HCFC-22 through 1 scale height in the atmosphere assuming an eddy diffusion coefficient of $K = 1 \times 10^5$ cm^2 s^{-1}.

(e) Discuss the expected zonal and vertical distributions of HCFC-22 in the atmosphere.

P1.12: In the middle and upper stratosphere, ozone concentrations are maintained at roughly steady values by a series of chemical reactions. Assume that at a temperature of 220 K the ozone mole fraction, x, is determined by the rate equation:

$$\frac{\mathrm{d}x}{\mathrm{d}t} = k_1 - k_2 x^2$$

where

$$k_1 = A \exp\left(\frac{300}{T}\right)$$

$$k_2 = B \exp\left(-\frac{1100}{T}\right)$$

and A and B are constants with units of s^{-1}. Doubling the concentration of CO_2 in the atmosphere is predicted to cool the middle stratosphere by about 2 K. What fractional change in x would you expect from this temperature change? You may assume that at each temperature the ozone mole fraction is in steady state.

Chapter 2

Radiation in the Atmosphere

Introduction

In the previous chapter, we highlighted the fact that the solar radiation that reaches the Earth's surface is absorbed and then re-emitted, with local variations in heating rate giving rise to the dynamic motion (transport) of air and contributing to general atmospheric circulation. In this chapter, we will highlight the role that O_3 and O_2 play in screening the Earth's surface from harmful UV radiation, which can damage animal and plant life. Absorption of UV and infrared (IR) wavelengths by the atmosphere results in the majority of the solar radiation that reaches the Earth's surface lying in the visible region of the electromagnetic spectrum. This radiation is absorbed and then re-emitted, and because the Earth is far colder than the Sun, the re-emitted radiation no longer lies in the visible, but rather in the IR region of the spectrum. Many atmospheric trace gases are IR-active and can absorb such terrestrial radiation. Trapping of heat within the atmosphere in this way leads to an increase in the temperature at the Earth's surface, a phenomenon commonly known as the greenhouse effect. In the following, we will employ simple radiative transfer models to explain the greenhouse effect and will consider a number of factors that influence the radiative balance within the atmosphere. In particular, we will consider the effect of increasing CO_2 levels.

2.1 The Solar Spectrum and Its Attenuation by the Atmosphere

2.1.1 The Black Body Model

The Sun's emission spectrum, prior to absorption by the Earth's atmosphere, is approximately that of a *black body* at a temperature of 6000 K. To understand the behaviour of a black body, consider an evacuated container whose walls are maintained at a constant and uniform temperature T. If a small object is suspended within the container and is in thermal equilibrium with the container, then the object will radiate as much energy as it absorbs. An object that is a good emitter at a particular wavelength, λ, is therefore also an equally good absorber at the same wavelength. A consequence of this behaviour is that the spectrum of radiation within the container at a given temperature is *independent* of the materials from which the container is made. If this were not the case, then energy would be transferred between objects at the same temperature. A black body at temperature T emits an amount of energy E per unit surface area per second, as given by the *Stefan–Boltzmann law*:

$$E = \sigma T^4 \qquad (2.1)$$

where $\sigma = 5.7 \times 10^{-8}\,\mathrm{W\,m^{-2}\,K^{-4}}$ is the Stefan–Boltzmann constant.

If an aperture with dimensions very much smaller than the linear dimensions of the container is left open then the black body acts as a source of radiation and some example emission spectra are shown in Figure 2.1.[1] The aperture acts as both a perfect emitter, transmitting all of the radiation coming from within the container, and as a perfect absorber, absorbing all of the light incident on the aperture. A black body therefore represents a set of standard, "perfect" emission characteristics with which to compare other emitters. The spectral *emittance*, ε_λ, of a surface is defined as the fraction of radiation emitted at a given λ compared to that emitted by a black body. Similarly, the spectral *absorptance*, α_λ, is the fraction of radiation that a surface absorbs relative to a black body at the same wavelength. From the discussion above it is clear that $\varepsilon_\lambda = \alpha_\lambda$, a result

---

[1]It can be seen that the most probable frequency in the black body emission spectrum increases with increasing temperature of the black body. This behaviour is encapsulated by the *Wien displacement law*, $\lambda_{\max}T = hc/5k_B = 2.89 \times 10^{-3}$ m K where $\lambda_{\max}$ is the most probably wavelength of the black body emission.

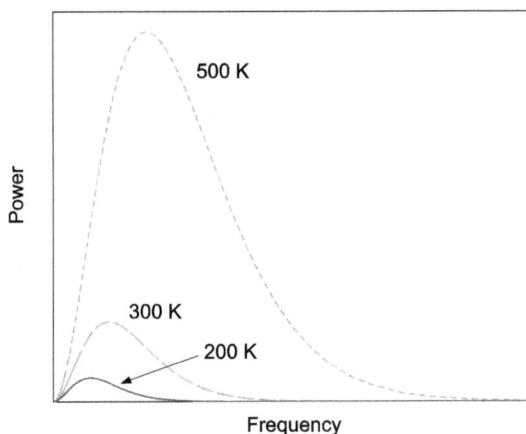

Figure 2.1 Example black body emission spectra which are a function of temperature only.

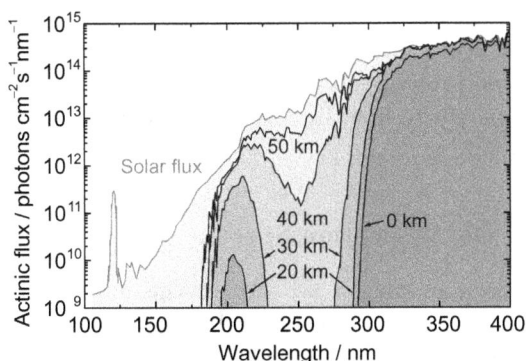

Figure 2.2 The actinic flux at a selection of altitudes.

Source: Adapted from *Chemical Kinetics and Photochemical Data for Use in Stratospheric Modelling* by W.B. DeMore *et al.*, JPL Publication, 97–104 (1997).

known as *Kirchoff's law*. Emittance and absorptance will be used later in this chapter when discussing the greenhouse effect.

2.1.2 The Solar Spectrum

Figure 2.2 shows a comparison between the solar spectrum and the spectrum of light reaching the surface of the Earth, expressed in terms of *actinic flux*. Actinic flux is defined by the International union of pure and applied chemistry (IUPAC) as "the quantity of light available to molecules at a particular point in the atmosphere, which, on absorption, drives photochemical processes in the atmosphere". Actinic flux is calculated by integrating the spectral radiance, $I(\lambda, \theta, \phi)$, over all incident directions (θ, ϕ) of the light. As will be discussed in more detail in the next section, the presence of O_2 and O_3 in the upper atmosphere (the stratosphere and above) ensures that virtually no radiation with $\lambda < 300$ nm can reach the Earth's surface (see also Sections 2.5 and 3.1). The strong absorption of UV radiation at high altitudes is also responsible for the marked variation with altitude of the rate constant for O_2 photolysis. This is in stark contrast to rate constants for photolysis processes occurring at longer wavelengths, for example NO_2 photolysis, which occurs at wavelengths below 410 nm and initiates formation of ozone in the troposphere (see Section 4.4). For these processes, suitably energetic radiation is available at all altitudes, and the rate constants show little variation with altitude.

In addition to UV light, IR radiation is also strongly absorbed by the atmosphere, in this case through excitation of vibrational transitions in a variety of trace molecular species, with water turning out to be particularly important in this regard. In comparison, the atmosphere is relatively transparent to light in both the visible and microwave regions of the electromagnetic spectrum.

2.1.3 O_3 and O_2 Absorption Spectra

Ozone and molecular oxygen are two of the most important absorbing species in the Earth's atmosphere, and we shall therefore consider their absorption spectra in some detail. Figure 2.3 shows the wavelength-dependent UV absorption cross-sections, $\sigma(\lambda)$, for both molecules.

2.1.3.1 *O_2 absorption*

The longest UV λ region of the O_2 absorption spectrum, spanning the range from $\sim$200 to 242 nm, is known as the *Herzberg continuum*, and comprises electronic transitions from the ground state $X^3\Sigma_g^-$ to the higher states $A^3\Sigma_u^+$, $A'^3\Delta_u$, and $c^1\Sigma_u^-$. Although all three transitions are formally forbidden, the molecular oxygen concentration in the middle and upper atmosphere is sufficiently high that the transition is easily observed. Absorption of radiation in this range leads to the dissociation of O_2 to produce ground state oxygen atoms, $O(^3P)$; 242 nm is the

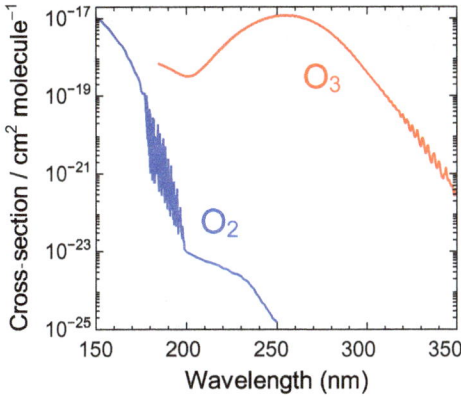

Figure 2.3   The UV absorption cross-sections of O_3 (red line) and O_2 (blue filled).

Source: The data are taken from J.H. Seinfeld and S.N. Pandis, *Atmospheric Chemistry and Physics: From Air Pollution to Climate Change*, Wiley (2006) (O_2) and L.T. Molina and M.J. Molina, *J. Geophys. Res.* **91** 14501 (1986) (O_3).

longest wavelength at which O_2 can be photolysed, and defines the lowest photon energy at which the photochemical cycle leading to the production of ozone can be initiated. The production of ozone will be considered in detail in Section 3.1. At shorter wavelengths, below 200 nm, lie the *Schumann–Runge* bands and continuum, which correspond to the fully-allowed $B^3\Sigma_u^- \leftarrow X^3\Sigma_g^-$ transition. Again, absorption leads to dissociation but now it is possible to form electronically-excited oxygen atoms, $O(^1D)$. Shorter wavelength Schumann–Runge absorption is important only at high altitudes, such as in the mesosphere and upper stratosphere. At lower altitudes within the stratosphere, where the Schumann–Runge wavelengths have already been removed, the longer-wavelength Herzberg absorption is more important.

2.1.3.2 O_3 absorption

Ozone exhibits strong UV absorption at wavelengths up to *ca.* 310 nm within the *Hartley band* ($^1B_2 \leftarrow X^1A_1$), with weaker absorption in the range 320–360 nm via the *Huggins band*.[2] Absorption in both of these bands induces dissociation and can lead to the formation of the highly reactive

---

[2]Interestingly, the Huggins band is due to a transition to the C_s wells of the B state, i.e. the same state responsible for the Hartley band. For more details see Z.-W. Qu *et al.*, *J. Chem. Phys.* **120** 6811 (2004).

electronically-excited $O(^1D)$ atom, the reactions of which will be highlighted in Chapters 3 and 4. O_3 also absorbs radiation in the visible and near-IR region via the *Chappuis band* which extends over the range 400–850 nm. While the maximum absorption cross-section for the Chappuis band is some three order of magnitude smaller than that for the Hartley band, we note that this band is still important in atmospheric chemistry as it occurs near the maximum of the solar spectrum and therefore drives absorption in the troposphere and lower stratosphere. In this case, the absorption of visible radiation leads to photodissociation to produce both atomic and molecular products in their ground electronic states, $O(^3P)$ and O_2 $(X^3\Sigma_g^-)$, respectively.

2.2 Theory of Absorption and Scattering by Atmospheric Gases and Particles

Having considered the spectrum of solar radiation incident on the Earth's atmosphere and transmitted through the atmosphere to the Earth's surface, and having identified some of the major absorbing species, we now take a more general view of the physics underlying absorption and scattering processes within the Earth's atmosphere.

2.2.1 Absorption and Scattering by Gases

2.2.1.1 *Refractive index*

An electromagnetic wave propagating through vacuum travels at the speed of light, $c = 2.99792458 \times 10^8 \, \mathrm{m\,s^{-1}}$. In any other medium, the propagation speed of the light is reduced by a factor equal to the (wavelength-dependent) refractive index, n, of the medium, i.e. $v = c/n$. The refractive index is often recast in terms of the relative electric permittivity, ε_r and the relative magnetic permeability, μ_r, of the medium:

$$n = (\varepsilon_r \mu_r)^{1/2} \tag{2.2}$$

For the Earth's atmosphere, μ_r can be taken as unity, and therefore we focus only on ε_r. For a uniform homogenous medium that does not absorb radiation, ε_r must be a real number. However, if the medium does absorb, ε_r must instead be represented as a complex number, typically in the form $\varepsilon_r = \varepsilon' - i\varepsilon''$. In this case n must also be complex, and we therefore define

$n = n' - in''$, where n' and n'' are the real and imaginary parts of the refractive index, respectively.[3]

2.2.1.2 *Absorption*

Consider an x-polarised wave propagating in the z-direction. The wave may be described mathematically by the expression

$$E_x = E_0 \exp\left(i(\omega t - kz)\right) \qquad (2.3)$$

where E_0 is the electric field amplitude, ω is the angular frequency, and k is the wavevector, whose magnitude is $k = 2\pi/\lambda = n\omega/c$. Absorption causes the amplitude of the wave to decrease exponentially with propagation distance z, as illustrated in Figure 2.4. Mathematically, the waveform becomes

$$E_0 \exp\left(-\omega n'' z/c\right) \exp\left(i\omega(t - n'z/c)\right) \qquad (2.4)$$

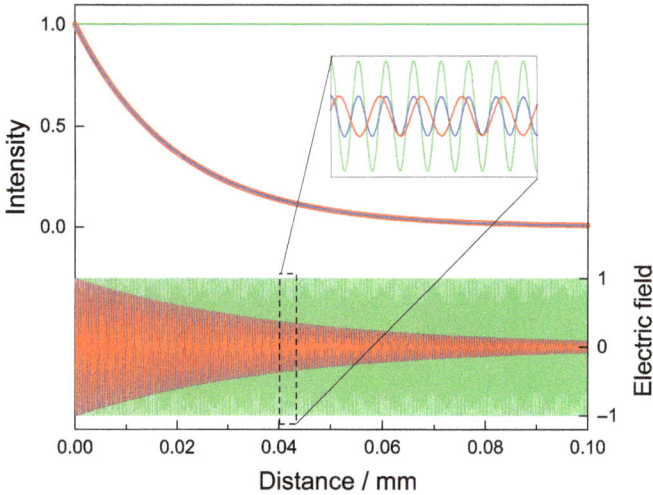

Figure 2.4 Normalised electric field amplitude (bottom) and intensity (top) with $\lambda = 500\,\text{nm}$ as a function of propagation distance for a series of refractive indices: $1.5 + 0.001i$ (blue); $1.01 + 0.001i$ (red); $1.5 + 0.000i$ (green). The inset shows details of the electric field oscillations.

---

[3]n'' is also known as the *extinction coefficient* as it quantifies the degree of attenuation when the electromagnetic wave propagates through a material.

Usually we are interested in the intensity of light, I, rather than the electric field. The intensity is proportional to the square modulus of the electric field, and therefore

$$I(z) = I_0 \exp\left(-\frac{2\omega n'' z}{c}\right) \tag{2.5}$$

where I_0 is the light intensity incident on the medium. This expression, which we first met in a slightly different guise in Chapter 1 in the first Volume of this series,[4] is known as the *Beer–Lambert law*. In that case, the Beer–Lambert law was used to relate spectral line intensities to the number density of absorbing species, i.e. the fraction of light transmitted through a sample of gas at a particular wavelength, λ, $I(\lambda)$, is given by

$$I(z) = I_0 \exp\left(-\sigma_a n_{mol} z\right) \tag{2.6}$$

where I_0 is the incident intensity of light at wavelength λ, n_{mol} is the number density of the gas, σ_a is the (molecule-specific) *absorption cross-section* at the wavelength of interest, and z is the path length of the light through the sample. By comparing Equation (2.5) with Equation (2.6), we see that the following equalities must hold:

$$\frac{2\omega n''}{c} = n_{mol}\sigma_a = \kappa_a \tag{2.7}$$

where κ_a is known as the *absorption coefficient*, or absorption per unit path length. The subscripts on σ_a and κ_a denote absorption, as distinct from the equivalent quantities for scattering which we will meet in the next section. With these substitutions, the Beer–Lambert law then takes the form given below.

$$I(z) = I_0 \exp\left(-\kappa_a z\right) \tag{2.8}$$

The product $\kappa_a z$ is commonly known as the *optical thickness*, often given the symbol χ. In terms of quantum mechanics, the absorption cross-section σ_a is proportional to the square of the transition dipole moment integral, which we met earlier in Section 5.3.4. of the first volume of this text. Since molecules only absorb light at characteristic frequencies, the quantities n', σ_a, κ_a, and χ all vary strongly with frequency or wavelength.

---

[4] *Astrochemistry: From the Big Bang to the Present Day* by C. Vallance.

It should also be noted that while spectroscopists typically use σ_a within the Beer–Lambert law, many texts on atmospheric physics use the formalism whereby Equation (2.8) is recast as

$$I(z) = I_0 \exp\left(-k_a \rho_a z\right) \tag{2.9}$$

where k_a is the *mass-weighted absorption cross-section* (m^2 kg^{-1}) and the *mass density* of the absorber, ρ_a, (kg of absorber per m^3 of air) is used rather than the number density, n_{mol}. Both formalisms will be used at appropriate points during the rest of this text.

2.2.1.3 *Scattering*

Molecules not only absorb radiation but can also *scatter* radiation, changing the direction of propagation without affecting the energy of the radiation. *Rayleigh scattering* occurs when the scatterer has dimensions much smaller than the wavelength of the radiation, for example scattering of light from molecules. Rayleigh scattering is a result of molecular electric polarisability. The oscillating electric field of the light interacts with the molecular charge distribution to induce an oscillating dipole with the same oscillation frequency as the incident radiation. The molecule then becomes a small radiating dipole, whose radiation is observed as scattered light as depicted in Figure 2.5. The cross-section, σ_s, for Rayleigh scattering is given by

$$\sigma_s = \frac{128\pi^5 a^6}{3\lambda^4} \tag{2.10}$$

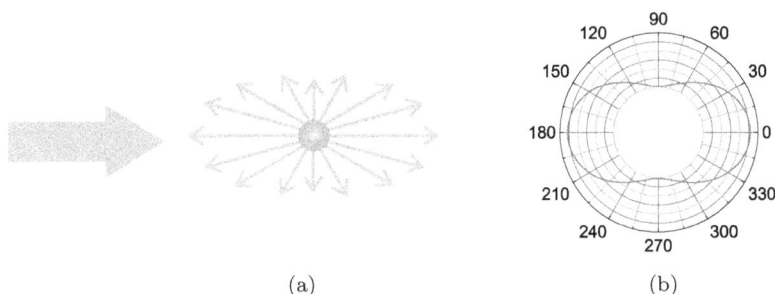

(a) (b)

Figure 2.5 (a) Schematic of Rayleigh scattering and (b) Rayleigh scattering phase function (linear scale).

where a is the effective radius of the molecule.[5] By analogy with the previous discussion, the *scattering coefficient*, κ_s, is $n_{mol}\sigma_s$, and the intensity of a light beam transmitted through a homogeneous sample of length z is

$$I(z) = I_0 \exp(-\kappa_s z) \tag{2.11}$$

To investigate the influence of Rayleigh scattering on the spectrum of light transmitted to the Earth's surface, consider scattering by molecules in air. Taking the effective radius of an 'air molecule'[6] to be 0.15 nm (similar to a bond length), a scattering cross-section of order 10^{-30} m^2 is calculated for red light. The column amount through the atmosphere is *ca.* 2×10^{29} molecules m^{-2}, and hence the optical thickness for scattering through the atmosphere is 0.2 for red light ($\lambda \sim 650$ nm), and 0.8 for blue light ($\lambda \sim 450$ nm). This preferential scattering of short-wavelength light is the reason that the clear sky appears blue, as blue light is more likely to be scattered in all directions, including towards the surface, than red light. At sunset and sunrise, when the Sun is close to the horizon as viewed from the Earth's surface, the light appears much redder than usual. This effect can also be explained by Rayleigh scattering: the path length through the atmosphere is now much longer, and much of the blue light has been scattered out of the line of sight to the Sun, leaving behind the less-efficiently-scattered longer-wavelength (redder) light.

Although σ_s allows calculation of the fraction of the initial radiation scattered by the particle, it says nothing about the spatial distribution of the scattered intensity. This latter information is encapsulated within the *phase function*, $p(\cos\Theta)$, where Θ is the angle through which the radiation has been deflected. For Rayleigh scattering, the phase function is given by

$$p(\cos\Theta) = \frac{3}{4}\left(1 + \cos^2\Theta\right) \tag{2.12}$$

such that scattering is maximised in the forward ($\Theta = 0$) and backward ($\Theta = \pi$) directions, and minimised for scattering through $\pi/2$.

---

[5] Molecules are not spherical and so the induced dipole moment is not necessarily in the direction of the applied electromagnetic field. The scattering is necessarily averaged over all molecular orientations and the effective radius is a simple way of accounting for this averaging.

[6] The properties of an "air molecule" are the appropriately weighted averages over the relevant properties for N_2, O_2, etc.

2.2.2 Absorption and Scattering by Particles

As discussed in the previous section, the effectiveness with which a molecule absorbs and scatters radiation is quantified by the cross-sections σ_a and σ_s. The presence of aerosol particles (solid and liquid) can also have a significant effect on the propagation of radiation within the atmosphere. We define absorption and scattering cross-sections for such particles in terms of the geometrical area πr^2 that the particle (of radius r) presents to incoming radiation:

$$\sigma_a = \pi r^2 Q_a$$
$$\sigma_s = \pi r^2 Q_s \tag{2.13}$$

where Q_a and Q_s are dimensionless quantities known as the *absorption* and *scattering efficiency*, respectively. The combined effect, $Q_a + Q_s$, of absorption and scattering is referred to as the *attenuation* or *extinction*. The absorption and scattering efficiencies are related to the attenuation coefficient, κ, introduced for molecular absorption and scattering in Equation (2.7), by

$$\kappa = n_{\text{particle}}(\sigma_a + \sigma_s)$$
$$= \pi r^2 n_{\text{particle}}(Q_a + Q_s) \tag{2.14}$$

The values of Q_a and Q_s, and therefore the value of κ, are determined by the size of the particle relative to the wavelength of the radiation, and also by its refractive index. In general, the wavelength dependence of κ follows the functional form

$$\kappa = \kappa_0 \lambda^{-m} \tag{2.15}$$

where κ_0 is a constant and m is known as the *Ångstrom exponent*. It is useful at this point to introduce the (dimensionless) *size parameter*, $x = 2\pi r/\lambda$. Rayleigh scattering (from small particles, such that $x \ll 1$) corresponds to the case for which $m = 4$. When $x \approx 1$, the particles exhibit a different type of scattering known as *Mie scattering* with $m = 0.2$–2, while scattering from very large particles yields a scattering cross-section equal to the geometrical cross-section, with $m = 0$.

Examples of $Q_a(x)$ and $Q_s(x)$ are shown in Figure 2.6, for the cases of weakly and strongly absorbing particles with complex refractive indices $n = 1.5+0.0001i$ and $n = 1.74+0.44i$, respectively. For small values of x, Q_s is proportional to x^4, while Q_a is proportional to x. When $r \sim \lambda$, Q_s tends towards its maximum value (with some oscillations) for both particles. In

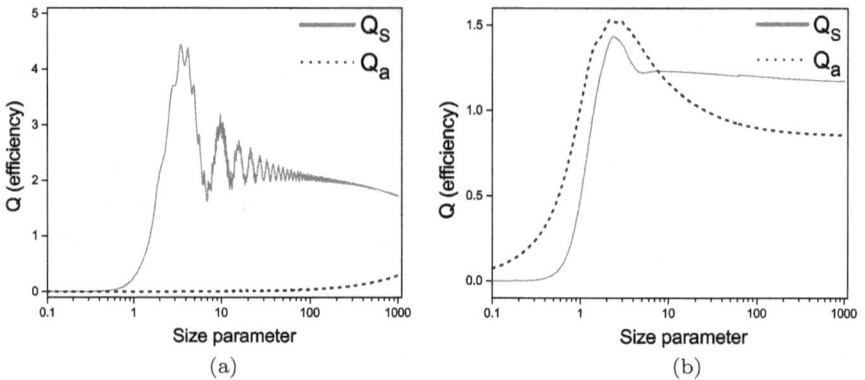

(a)

(b)

Figure 2.6   Absorption and scattering efficiencies as a function of size parameter for (a) a weakly absorbing particle with $n = 1.5 + 0.0001i$ and (b) a strongly absorbing particle with $n = 1.74 + 0.44i$. Both particles are situated in a medium with $n = 1$.

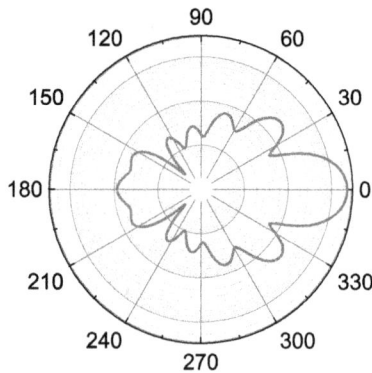

Figure 2.7   Phase function for Mie scattering. The particle has a radius of $1\,\mu m$ and the incident light has a wavelength of $500\,nm$. Note the logarithmic scale.

the case of the strongly absorbing particle, Q_a is also a maximum when $r \sim \lambda$, while for the weak absorber Q_a is close to zero across much of the size parameter range (as expected). Finally, for large particles, unless the refractive index has significant λ dependence, the scattering and absorption are independent of x. We note that such a strongly absorbing particle is a good model for *soot/black carbon* — see Section 5.4.

The phase function for Rayleigh scattering was given in Equation (2.12). The phase function for Mie scattering is more complicated, typically exhibiting a number of angular nodes of order x. Figure 2.7 shows the phase function for a particle of radius 1 μm illuminated by visible radiation with $\lambda = 500$ nm. In comparison to Rayleigh scattering, Mie scattering exhibits a much larger degree of spatial anisotropy, with short wavelengths being strongly forward scattered.

For typical atmospheric aerosols, m lies in the range 0.2–2. Marine aerosols usually exhibit the largest particle sizes (see Section 5.1) and hence the smallest values of m. For most other tropospheric aerosols m is *ca.* 1. The attenuation coefficient, κ, for visible radiation at sea level typically lies in the range 0.05–0.5 km^{-1} while the vertical optical thickness of tropospheric aerosol is in the range 0.1–1; these values highlight the importance of aerosols in determining the radiative properties of the atmosphere. Furthermore, it should be noted that the attenuation of visible and IR radiation by aerosol is generally dominated by scattering rather than absorption and, as shown above, κ is both size and wavelength dependent.

The large water droplets found in low altitude clouds typically have radii, r, of the order of tens of μm. In this case $(x \gg 1)$, the scattering cross-section σ_s is roughly equal to the geometrical cross-sectional area of the droplet, and the scattering coefficient is $\kappa_s = n_d \pi r^2$ where n_d is the number of droplets per unit volume of air. The mass density, ρ, of liquid water *in the cloud* is

$$\rho = n_d \frac{4}{3}\pi r^3 \rho_l \tag{2.16}$$

where ρ_l is the density of liquid water. We can rearrange this expression to find n_{droplet}, yielding a new expression for the scattering coefficient:

$$\kappa_s = \frac{3\rho}{4r\rho_l} \tag{2.17}$$

The droplet radius is not strongly dependent on mass density, and therefore κ_s is directly proportional to the mass density of liquid water in the cloud. Within clouds, ρ is generally in the range from 10^{-4} to 10^{-2} kg m^{-3}, yielding κ_s values in the range 1–100 km^{-1}. Cloud layers are typically of the order of 1 km thick, and so all but the thinnest clouds are optically thick or *opaque*. The optical properties of aerosol and clouds will be considered further, both in this chapter and in Chapter 5.

2.2.3 Calculation of Photolysis Rates

The rate at which molecules are photolysed within the atmosphere depends directly on the available intensity of solar radiation of sufficiently short wavelength. Now that we have considered the various factors determining the intensity of solar radiation reaching various altitudes within the atmosphere, we can develop expressions for calculating photolysis rates at any altitude.

Consider a generic photolysis process

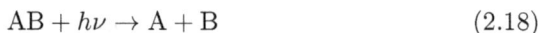

$$AB + h\nu \rightarrow A + B \tag{2.18}$$

The rate constant, J, for such a process is given by

$$J = \int \phi_\nu I_\nu \sigma_\nu d\nu \tag{2.19}$$

where the integral is over all frequencies, ν, at which the species AB absorbs radiation; ϕ_ν, I_ν, and σ_ν are the photolysis quantum yield (probability of dissociation on absorption of a photon), actinic flux of radiation, and molecular absorption cross-section at frequency ν. The quantum yield is zero at photon energies below the dissociation threshold, often rising to near unity above the threshold. Values significantly less than unity over this energy range indicate the presence of one or more competing processes that consume the excited state AB molecules formed following photon absorption.

As discussed in Section 2.2, the actinic flux varies across a given layer of the atmosphere according to the Beer–Lambert law,

$$I_\nu(z) = I_{\infty,\nu} \exp(-\chi) \tag{2.20}$$

where $I_{\infty,\nu}$ is the intensity of radiation with frequency ν at the top of the layer, and χ is the optical thickness within the layer, given by

$$\chi = \int_z^\infty \Sigma_j[j](z)\sigma_{j,\nu} dz' \tag{2.21}$$

Here, the integral is over the (not necessarily vertical) geometrical path z' through the atmosphere, and the sum is over all species j that absorb at frequency ν, and $[j]$ is the concentration of species j (also a function of altitude z). The optical depth (and hence actinic flux) depends upon all absorbing species at the frequency of interest, but to a very good approximation, only the absorbers O_2 and O_3 need be considered in calculating $I_\nu(z)$. Remembering that O_2 only absorbs at $\lambda < 242\,nm$ and O_3 only absorbs (strongly) at $\lambda < 310\,nm$, the optical thickness at a given

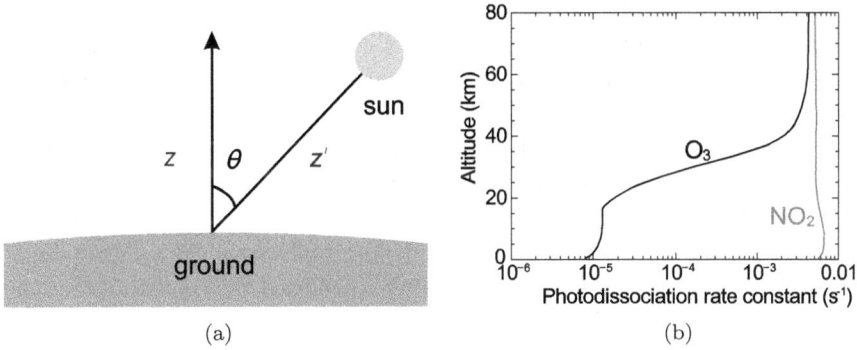

(a)   (b)

Figure 2.8 (a) Definition of the solar zenith angle and (b) photolysis rate constants for O_3 and NO_2 as a function of altitude.

Source: Adapted from *Photochemistry of Planetary Atmospheres* by Y.L. Yung and W.B. DeMore, Oxford University Press (1999).

frequency can be calculated from

$$\chi(\nu) = \int_z^\infty \left([O_2](z)\sigma_{O_2,\nu} + [O_3](z)\sigma_{O_3,\nu}\right) dz'$$

where z' is the path length as depicted in Figure 2.8(a). It should be noted that the concentrations are a function of altitude, z, and *not* path length, z'. The path length varies with the *solar zenith angle* θ (measured from vertical) with $z' = z/(\cos\theta)$. The path length, and therefore the optical depth, increases rapidly as θ approaches 90°, yielding considerably lower light intensities at high latitudes or at sunrise/sunset.[7] For vertical paths, the solar zenith angle is zero, and $z' = z$. Under these conditions, the optical thickness is

$$\chi = U_{O_3,z}\sigma_{O_3,\nu} + U_{O_2,z}\sigma_{O_2,\nu} \qquad (2.22)$$

with $U_{j,z}$ the overhead column amount of species j from altitude z to the top of the atmosphere (TOA). The rate constant for photolysis of AB is then given by

$$J = \sum_\nu \phi_\nu \sigma_\nu I_{\infty,\nu} \exp\left(-(U_{O_3,z}\sigma_{O_3,\nu} + U_{O_2,z}\sigma_{O_2,\nu})z\right) \qquad (2.23)$$

In summary, photolysis rate constants (and therefore photolysis rates) depend upon a number of factors, including the frequencies at which the

---

[7]A more subtle point is that while the optical thickness is increased by a factor of $1/(\cos\theta)$ for angles up to *ca.* 75°, at larger angles it is necessary to take the Earth's curvature into account.

species of interest absorbs and its absorption cross-section and photolysis quantum yield at these frequencies; and the incident radiation intensity, which varies with atmospheric composition, phase within a solar cycle, and solar zenith angle. For the most accurate predictions of photolysis rates, we should consider all species within the atmosphere that absorb at the frequencies of interest, their overhead column amounts at the altitude of interest and their absorption cross-sections, as well as molecular scattering, and scattering from clouds at lower altitudes. As an example of these considerations, Figure 2.8(b) shows calculated rate constants for the photolysis of O_3 and NO_2 as a function of altitude.

2.3 The Greenhouse Effect

As described earlier, the surface of the Earth absorbs incoming solar radiation and subsequently acts as a pseudo-black body emitter. As the Earth is far colder than the Sun, the peak emission wavelength of the terrestrial radiation lies in the IR at around 15 μm. For comparison, the peak emission wavelength for the Sun lies in the visible at around 500 nm. The black body spectra of the Sun and the Earth are shown in Figure 2.9. Much of the outgoing IR radiation emitted from the Earth is absorbed by trace gases in the atmosphere, trapping the energy within the atmosphere and leading to increased surface warming, a phenomenon known as the greenhouse

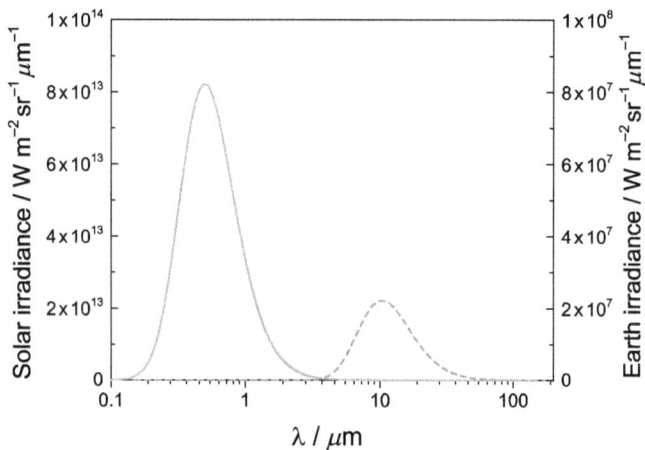

Figure 2.9   The black body emission spectrum of the Sun (solid line) compared to the Earth (dotted line). Earth is assumed to have $T = 280$ K.

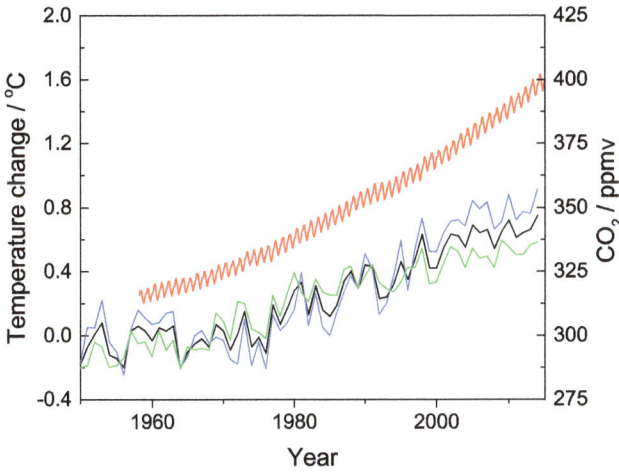

Figure 2.10 Atmospheric levels of CO_2 (red) measured at Mauna Loa plotted with surface temperature change averaged over the Northern hemisphere (blue), Southern hemisphere (green), and globe (black) as a function of time. The temperature change is relative to the average surface temperature from 1951 to 1980. [CO_2 data: Dr. Pieter Tans, NOAA/ESRL (www.esrl.noaa.gov/gmd/ccgg/trends/) and Dr. Ralph Keeling, Scripps Institution of Oceanography (scrippsco2.ucsd.edu/); temperature data from Goddard Institute for Space Studies, NASA.]

effect. Carbon dioxide is an important *greenhouse gas* (GHG), and rising levels of CO_2 in the atmosphere are synonymous with the phenomenon of *global warming*. Figure 2.10 shows the increasing levels of CO_2 over the last 50 or so years alongside the increased surface temperature; data such as this provide evidence for the role of anthropogenic activities in augmenting the atmosphere's inherent (natural) greenhouse effect. In the following, we will discuss the underlying physics that determines the degree of surface warming, whilst also noting that such warming will also lead to stratospheric cooling.[8]

With regard to stratospheric cooling, consider a situation in which both the troposphere and the surface warms by a small amount ΔT, as shown in Figure 2.11. Tropospheric air is well mixed, and if the troposphere

---

[8]The atmospheric CO_2 concentration exhibits strong seasonal oscillations in addition to the rise noted over the last century. CO_2 levels decline when the terrestrial vegetation of the Northern hemisphere awakens from the dormancy of winter and begins to grow in the spring, thereby extracting large quantities of CO_2 from the air. The levels then rise again in the autumn and winter, when much of the biomass produced over the summer dies and decomposes, releasing large quantities of CO_2 back to the atmosphere.

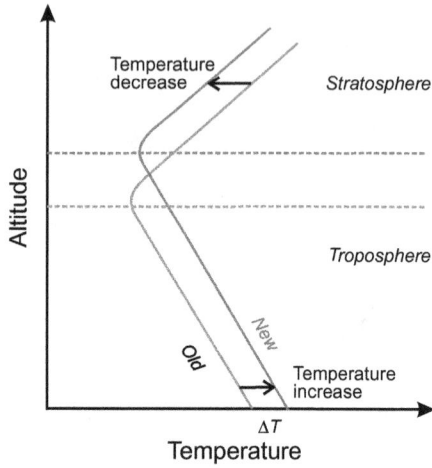

Figure 2.11 A schematic of the stratosphere cooling which accompanies surface warming.

warms then the temperature profile will shift upwards as shown, with the adiabatic lapse rate remaining constant. This immediately leads to *cooling* of the stratosphere so that energy balance is maintained. As shown in Section 1.3.3, absorption of UV radiation by O_3 heats the stratosphere, while CO_2 and H_2O emission causes cooling, and these processes are normally in near-equilibrium. If the stratosphere is treated in terms of a homogeneous isothermal black body at temperature T_{strat} then its IR emission rate, E, can be calculated, assuming the Stefan–Boltzmann law (Equation (2.1)), and our earlier definition of the emittance (see Section 2.1.1), to be

$$E = \varepsilon\sigma T_{strat}^4 = (1-t)\sigma T_{strat}^4 \qquad (2.24)$$

where ε, is the emittance of the stratosphere and t, is its transmission. Addition of more CO_2 to the stratosphere leads to increased IR absorption, thereby decreasing t (and also increasing ε). Assuming that the solar input does not change and that energy balance must be maintained, then T_{strat} must fall; the resultant change in temperature can then influence atmospheric composition.

2.3.1 The Balanced Flux Model

The simplest model for estimating the temperature of the Earth, T_E, assumes that the Earth absorbs all of the solar radiation incident on it,

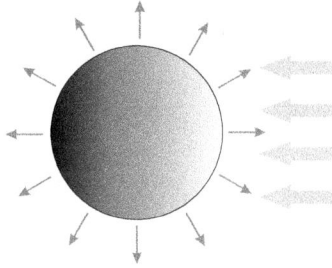

Figure 2.12 Solar light impinging on Earth (grey arrows) and long-wave radiation emitted by the planet (solid arrows).

and subsequently re-radiates it (see Figure 2.12). The solar flux, F_s, is $1370\,\mathrm{W\,m^{-2}}$, and so the total solar energy incident upon the Earth, which presents a cross-sectional area πR^2 to the incoming sunlight, with R the Earth's radius, is

$$E_{in} = \pi R^2 F_s \tag{2.25}$$

Assuming that the Earth acts as a black body with surface area $4\pi R^2$, and that its atmosphere is transparent to the emitted radiation, then the amount of re-radiated energy emitted by the Earth, E_{out}, is

$$E_{out} = 4\pi R^2 \sigma T_E^4 \tag{2.26}$$

Equating the incoming and outgoing energy yields a temperature for the Earth of $T_E = 280\,\mathrm{K}$. While this is not too far away from the true average surface temperature, the apparent success of the balanced flux model is entirely fortuitous, as the calculation omits two large and partly compensating effects. Firstly, the model neglects the Earth's *albedo*, A, which quantifies the fraction of incoming solar radiation that is reflected directly back into space, due to reflection at the Earth's surface and scattering by molecules and clouds in the atmosphere. The albedo can be highly surface dependent, as shown in Table 2.1, but a reasonable global average value is 0.3, i.e. 30% of incoming solar radiation is reflected back to space and not absorbed at the Earth's surface. Including the Earth's albedo in the calculation of E_{in} gives

$$E_{in} = (1 - A)\pi R^2 F_s \tag{2.27}$$

Energy balance then yields a value for T_E of 255K, which is significantly lower than the observed mean surface temperature. T_E from this calculation is commonly known as the *effective emitting temperature* of the Earth.

Table 2.1   A selection of albedos averaged over
the visible spectrum for various surface types.

Surface type	Albedo (fraction)
Liquid water	0.05–0.2
Snow	0.4–0.95
Clouds	0.2–0.9
Sea ice	0.25–0.4
Desert	0.2–0.4
Forest	0.1–0.25
Soil	0.05–0.2

As most readers will have realised already, given the discussion in the previous sections, the second shortcoming of our calculation is the assumption that the Earth's atmosphere is transparent to the IR radiation emitted from the Earth. The IR absorption cross-sections for a selection of important atmospheric trace gases are shown in Figure 2.13 and we see that many atmospheric species absorb in the wavelength range emitted by a $\sim$300 K black body such as the Earth. Far from being transparent, the atmosphere is strongly absorbing (optically thick), particularly over the Equator, where the partial pressure of water is greatest. Some of this absorbed radiation is then redirected back to the Earth's surface, such that E_{out} is lower than predicted in our simple calculation above.

Collectively, the trace gases that absorb within the black body spectrum emitted by the Earth are known as greenhouse gases (GHGs). The relative importance of each GHG depends both on its abundance and on its individual spectral properties (e.g. its absorption cross-section, spectral range, and overlap of its absorption spectrum with those of other species), as will be discussed in Section 2.3.3. For example, the 15 μm peak emission wavelength for the Earth corresponds to a spectral region where CO_2 is the principal absorber. We note that while water absorbs strongly across the relevant wavelength range, and plays a very important role in a phenomenon known as *radiative forcing* (see Section 2.4.2), it is not normally considered to be a GHG as its concentration depends strongly on atmospheric temperature, and therefore on the presence of CO_2 and other GHGs. Instead, water is considered to be a *feedback* species, reflecting changes in abundance of other GHGs. Absorption by GHGs leads to the breakdown of the simple balanced flux model, and more comprehensive models are required in order to obtain a satisfactory description of the radiation balance within the atmosphere. We shall develop one such model in the following section.

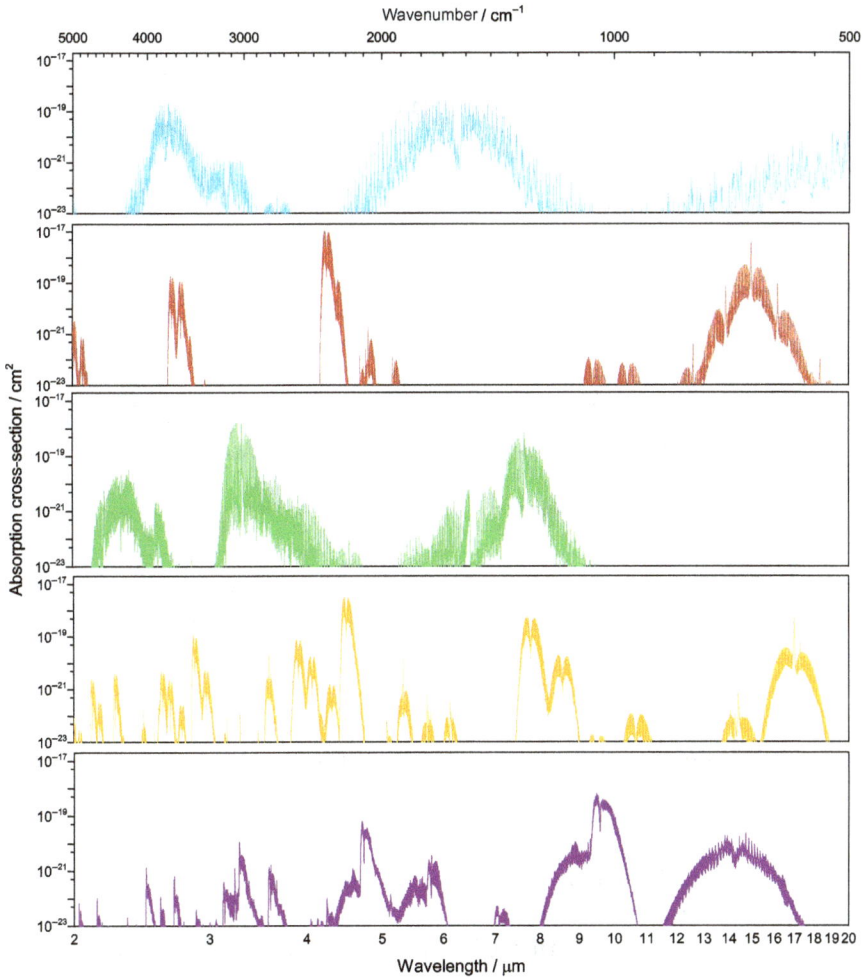

Figure 2.13   IR absorption cross-sections for important GHGs: H_2O (blue), CO_2 (red), CH_4 (green), N_2O (yellow), and O_3 (violet).

Source: Data from the **high-resolution tran**smission molecular absorption database (HITRAN) spectral database (https://www.cfa.harvard.edu/hitran/welcometop.html).

2.3.2 Single Layer Atmosphere Radiative Model

We now extend the simple energy balance model introduced above to an isothermal atmosphere having temperature T_{atm}. Consider Figure 2.14, in which t_{vis} and t_{IR} are the transmissions through the atmosphere at visible and IR wavelengths respectively. We start by calculating the average

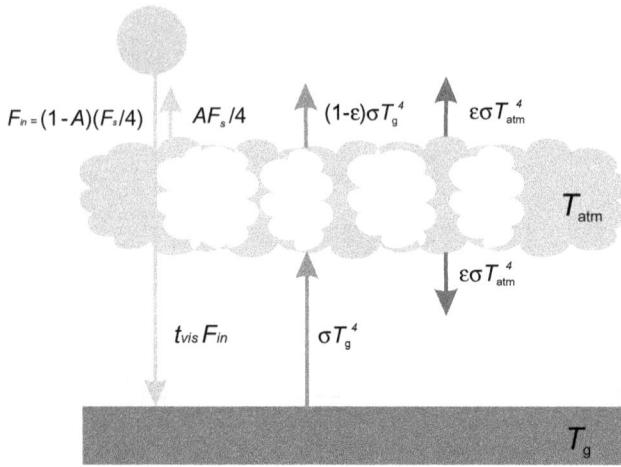

Figure 2.14 A schematic of the one-layer radiative model.

incoming radiation flux (energy absorbed or emitted per unit area) in the same way as previously, i.e. the global average incoming solar flux is

$$F_{\text{in}} = \frac{E_{\text{in}}}{4\pi R^2} = \frac{1-A}{4} F_s = 240 \text{ W m}^{-2} \tag{2.28}$$

The flux of outgoing radiation emitted from the Earth's surface is

$$F_{\text{out}} = \sigma T_g^4 \tag{2.29}$$

We now need to include in our model the flux of radiation emitted by the atmosphere and we do this by assuming that the emitted atmospheric flux, F_{atm}, can be modelled as that of an pseudo-black body with emittance ε (related to the IR transmission, t_{IR}, by $\varepsilon = 1 - t_{\text{IR}}$):

$$F_{\text{atm}} = \varepsilon \sigma T_{\text{atm}}^4 \tag{2.30}$$

Energy balance requires that the net flux at all altitudes is zero. Balancing the various fluxes therefore gives rise to the following two conditions:

(1) At the TOA, $F_{\text{in}} = F_{\text{atm}} + t_{\text{IR}} F_{\text{out}}$;
(2) At the ground, $t_{\text{vis}} F_{\text{in}} + F_{\text{atm}} = F_{\text{out}}$.

By eliminating F_{atm} from these two simultaneous equations, it follows that $(1 + t_{\text{vis}}) F_{\text{in}} = (1 + t_{\text{IR}}) F_{\text{out}}$. Substituting for the incoming and outgoing fluxes from Equations (2.28) and (2.29) and solving to find the surface temperature, T_g, gives

$$T_g = \left(\frac{(1-A)}{4\sigma} \frac{(1+t_{vis})F_{in}}{(1+t_{IR})} \right)^{1/4} \tag{2.31}$$

Taking values of $t_{vis} = 0.8$ and $t_{IR} = 0.1$, i.e. the atmosphere is a good transmitter of visible radiation but a strong absorber of IR, yields a surface temperature T_g of 288.1 K, which is consistent with our experience. The same one-layer model predicts the temperature of the atmosphere, T_{atm}, to be 250 K, which is also consistent with observation. Thus it can be concluded that the atmosphere exhibits a 'natural' greenhouse effect, in that its absorbance of IR radiation results in warming of the Earth to a temperature around 30 K higher than it would be otherwise. The one-layer model also predicts that the addition of more GHGs to the atmosphere, and the concomitant reduction in t_{IR}, results in higher temperatures. For example, if t_{IR} changes from 0.1 to 0.05 then T_g will increase by around 3.4 to 291.5 K. In practice, there are several feedback mechanisms operating within the atmosphere, and which the simple one-layer model neglects when it assumes that the state of the atmosphere remains constant except for a change in t_{IR}. For example, as the temperature increases, so does the water content in the atmosphere. This causes t_{IR} to reduce even more, further amplifying the temperature increase in an example of a *positive feedback* effect. We will consider the long-term response of the atmosphere to the increasing CO_2 levels it is experiencing at present in Section 2.4.2. Despite the success of the one-layer model, it should be emphasised that the model assumes the entire atmosphere to be in radiative equilibrium, and completely neglects any non-radiative energy transfer processes associated with fluid motion. In reality, non-radiative processes such as convection also contribute significantly to the energy balance, as we shall see in Section 2.4.1.

2.3.3 The Relative Effectiveness of Different GHGs

GHG absorption of mid-IR radiation induces transitions between vibrational energy levels. In order for a molecule to absorb or emit IR radiation, the vibrational motion must give rise to a *changing* electric dipole. As a consequence, the homonuclear diatomic molecules O_2 and N_2 (luckily for the inhabitants of Earth) display no mid-IR absorption. Many trace gases, however, have a number of vibrational modes that can be activated through

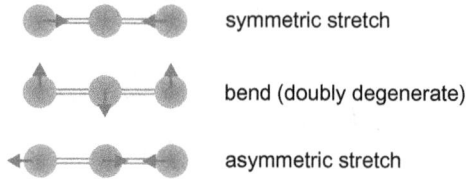

symmetric stretch

bend (doubly degenerate)

asymmetric stretch

Figure 2.15   The vibrational modes of CO_2.

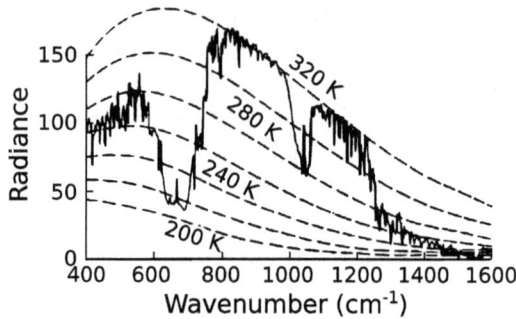

Figure 2.16   An example of a satellite measurement of the IR emission from the atmosphere above the Sahara desert. Radiance units: erg s^{-1} cm^{-2} ster^{-1}/cm^{-1}.

Source: Adapted from R.A. Hanel *et al.*, *J. Geophys. Res.* **77** 2629 (1972).

absorption of IR radiation as indicated by Figure 2.13. The linear molecule CO_2, for example, has four vibrational modes (see Figure 2.15), three of which are IR-active.

Figure 2.16 shows the IR emission escaping our atmosphere, as measured by a satellite positioned in space and viewing the Sahara desert. The spectral signatures of GHGs are compared to the black body emission spectra at various temperatures; the black body spectra are relatively unstructured and are denoted by the dashed lines. It is clear that H_2O acts as a strong GHG as it absorbs everywhere, cf. the 1400–1600 cm^{-1} region. In the atmospheric 'window' region between 10 and 12 μm (850–1000 cm^{-1}) where absorption by H_2O vapour is a minimum, and in the absence of clouds, the radiation reaching space from Earth originates mostly from the Earth's surface; the spectra are consistent with a black body at a temperature of 320 K. To longer wavelength the mean absorption coefficient of CO_2 gradually increases. As a result of repeated absorption and re-emission, the average altitude from which radiation is emitted increases, until at the most strongly absorbed wavelength — corresponding to the Q branch of the CO_2 ν_2 band near $667\,cm^{-1}$ ($15\,\mu m$) — the radiation escaping to space originates largely in the stratosphere with a local temperature of *ca.* $220\,K$. Measurements of IR spectra therefore allow *vertical temperature profiling* of the atmosphere to be carried out.

The relative effectiveness of a particular GHG *per molecule* depends strongly on the wavelengths at which the molecule absorbs. The most effective GHGs absorb close to the maximum of the terrestrial black body emission, absorb in many spectral regions, and in particular, absorb strongly in the atmospheric window region from 10 to $12\,\mu m$. However, GHGs cannot be considered in isolation. A potential GHG may satisfy all of the stated criteria, and still induce little, if any, greenhouse effect if the wavelengths at which it absorbs strongly have already been removed through absorption by much more abundant atmospheric gases such as CO_2 and water.

This last effect may be measured in terms of the *degree of saturation* at a particular wavelength, which depends on the optical depth, χ. At small χ (small total concentration of absorbing molecules and small absorption cross-sections), examination of the Beer–Lambert law reveals that the amount of absorbance depends linearly on the absorber concentration. At large χ (high total concentration) however, the absorbance is already high, and so adding another molecule only slightly increases the net absorption through increased absorption in the wings of a saturated band, e.g. doubling the concentration of CO_2 in the atmosphere will not double the amount of light absorbed, as the available light has already been absorbed by the existing CO_2. We will make this argument more quantitative in Section 2.3.4. Table 2.2 summarises the effectiveness of a selection of molecules that can act as GHGs, relative to that of CO_2. It should be noted that although CO_2 is not very effective on a per molecule basis, it is still the most important GHG overall due to its large abundance. CH_4, N_2O and CFCs are also important, as not only are they more effective per molecule than but CO_2 they are all long-lived species.

2.3.4 Absorption Lineshapes and Saturation

As highlighted in the previous section, CO_2 absorption in the IR is *saturated* in the sense that the absorbance is not directly proportional to the CO_2 concentration. We now explore this point further. IR spectra generally display thousands of individual lines associated with transitions between different rotational and vibrational levels, and these often overlap within

Table 2.2   Relative effectiveness per molecule of a selection of GHGs.

Species	Absorbance	Effectiveness
CO_2	Principal GHG; Absorbs strongly near maximum of black body curve; No pure rotational transitions; Strong mid-IR vibrational transitions.	1
H_2O	Strong GHG (but not normally considered so); Strong pure rotational transitions (far-IR); Strong mid-IR vibrational transition.	—
O_3	Weak pure rotational transitions (far-IR); Strong mid-IR vibrational transition.	~2
CH_4	No pure rotational transitions; Strong mid-IR vibrational transition.	20
N_2O	Weak pure rotational transitions (far-IR); Strong mid-IR vibrational transition.	200
CFCs	Strong mid-IR vibrational transition.	10,000

a vibrational band.[9] The structure of a vibrational band is affected by line broadening arising from a number of mechanisms, principally pressure broadening (also known as collisional or lifetime broadening)[10] and Doppler broadening, which are dependent on total pressure and temperature, respectively. The contributions from the various line-broadening mechanisms vary throughout the atmosphere. As an example, we consider a simple model in which a vibrational band consists of a large number of non-overlapping lines which are only broadened as a result of collisions. The model provides a

---

[9]There are many excellent texts on molecular spectroscopy. See for example, J.M. Hollas: *Modern Spectroscopy*, 4th edition, Wiley (2004). In addition, interested readers are encouraged to try out PGOPHER, a general purpose program for simulating and fitting rotational, vibrational and electronic spectra. The program is freely downloadable from the web site http://pgopher.chm.bris.ac.uk/ for Microsoft Windows, Apple Mac, and Linux users.

[10]Lifetime broadening results from the quantum-mechanical uncertainty principle relating energy and time. A quantum state with lifetime τ has an uncertainty in its energy given by $\Delta E \tau = h/4\pi$. Excited states of molecules generally have short lifetimes (even under collision-free conditions), and the resulting uncertainty in their energy leads to spectral lines having a finite linewidth known as the *natural linewidth*. Any process that quenches an excited state and reduces its lifetime, e.g. collisional quenching, leads to a corresponding increase in the energy uncertainty and therefore an increase in the spectral linewidth. This line-broadening mechanism is distinct from Doppler broadening, which results from the velocity component of the absorbing/emitting species along the propagation direction of the radiation.

reasonable description of the behaviour of CO_2 and water vapour bands in the middle/upper stratosphere.

The absorption cross-section, $\sigma(\bar{\nu})$, for a single pressure-broadened rovibrational transition with line centre at wavenumber $\bar{\nu}_0$, is given by

$$\sigma(\bar{\nu}) = \frac{s\gamma}{\pi\left[(\bar{\nu} - \bar{\nu}_0)^2 + \gamma^2\right]} \tag{2.32}$$

where $s = \int_0^\infty \sigma(\bar{\nu})\mathrm{d}\bar{\nu}$ is the *line strength* and $\gamma = (1/(2\pi c\Delta t))$ is the halfwidth (in cm^{-1}) of the line. Δt is the mean time between collisions in the gas, and varies over several orders of magnitude in the atmosphere due to its inverse dependence on pressure. The absorption cross-section is plotted as a function of wavenumber in Figure 2.17 for the cases of high and low pressure. Ignoring the temperature dependence of the line halfwidth γ, which is small by comparison with the effects of pressure broadening, we can write $\gamma(p) = \gamma_0(p/p_0)$, where γ_0 is the halfwidth at the reference pressure, p_0. Typically γ_0 lies in the range $0.5\text{--}1.5\,\mathrm{cm}^{-1}\,\mathrm{atm}^{-1}$.

We now consider the consequences of the lineshape function described above. The transmission $t(\bar{\nu})$ through a path of length l filled with a gas

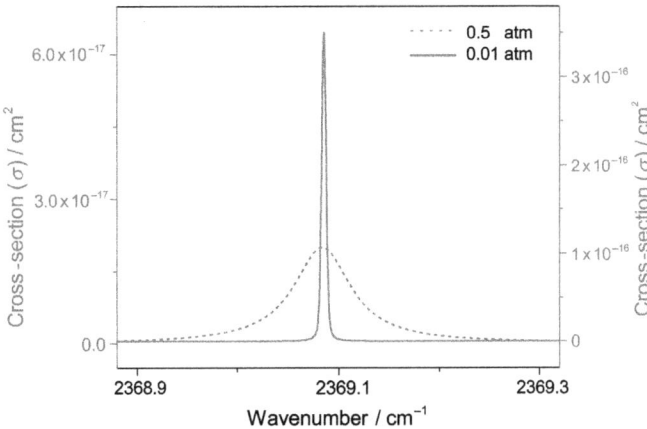

Figure 2.17 The absorption cross-section, $\sigma(\bar{\nu})$, for a CO_2 transition at two different air pressures ($T = 296\,\mathrm{K}$). At $0.01\,\mathrm{atm}$, the lineshape is Gaussian and is determined by Doppler broadening while at a total pressure of $0.5\,\mathrm{atm}$, pressure broadening is dominant and the lineshape is well described by a Lorentzian function. Note that while the absorption cross-section at line centre decreases with increasing pressure, the *integrated* cross-section is constant.

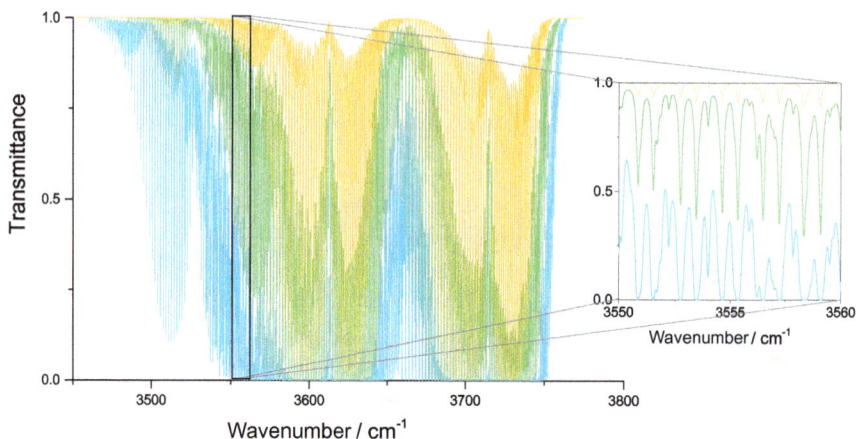

Figure 2.18   The transmission $t(\bar{\nu})$ through a 10 km path length for an atmospheric pressure sample containing 1 (yellow), 10 (green), and 100 (blue) ppmv of CO_2.

present at number density n_{mol} is given by the Beer–Lambert law

$$t(\bar{\nu}) = \exp\left(-\sigma(\bar{\nu})n_{mol}l\right) \qquad (2.33)$$

Figure 2.18 shows example transmission spectra for samples of CO_2 at mixing ratios of 1, 10, and 100 ppmv at atmospheric pressure, over a path length of 10 km. At very high values of the optical thickness, the sample absorbs all of the available radiation, and the transmission drops to zero. The line is then said to be *saturated*. In such cases, it becomes useful to define the *equivalent linewidth*, or *integrated absorptance*, W.

$$W = \int_0^\infty d\bar{\nu}\,(1 - t(\bar{\nu})) = \int_0^\infty d\bar{\nu}\,(1 - \exp\left(-\sigma(\bar{\nu})n_{mol}l\right)) \qquad (2.34)$$

There are two limiting cases in which this expression can be simplified, known as the *weak* and the *strong approximations*:

(1) The weak approximation applies when $\sigma(\bar{\nu})n_{mol}l \ll 1$. In this case, $t(\bar{\nu}) \sim 1 - \sigma(\bar{\nu})n_{mol}l$, and the integrated absorptance is $W = sn_{mol}l$;
(2) The strong approximation applies when the transmission is essentially zero near the line centre (in which case, the line is said to be black). In this case, the actual value of $\sigma(\bar{\nu})$ is not important, as long as it is sufficiently large enough for γ^2 to be omitted from the expression for the lineshape. Under these conditions $W = 2(s\gamma n_{mol}l)^{1/2}$, and the

integrated absorptance is no longer a linear function of concentration but is instead proportional to the square root of the concentration.

Finally, the average transmission, $\langle t \rangle$, over an absorption band of width $\Delta \bar{\nu}$ consisting of a number of non-overlapping lines is then given by

$$\langle t \rangle = 1 - \frac{\sum_i W_i}{\Delta \bar{\nu}} \tag{2.35}$$

where the sum is over the individual lines.

2.4 Global Energy Balance and Radiative Forcing

2.4.1 The Global Radiation and Energy Balance

In Section 2.3, energy balance was used to estimate the average surface temperature of the Earth. In the following, we further examine the global energy balance, with reference to Figure 2.19. The average incoming solar flux at the TOA is $343\,\mathrm{W\,m^{-2}}$ ($F_{\mathrm{in}} = F_{\mathrm{s}}/4$) and is known as the *insolation*. It should be noted that there are marked variations about this average,

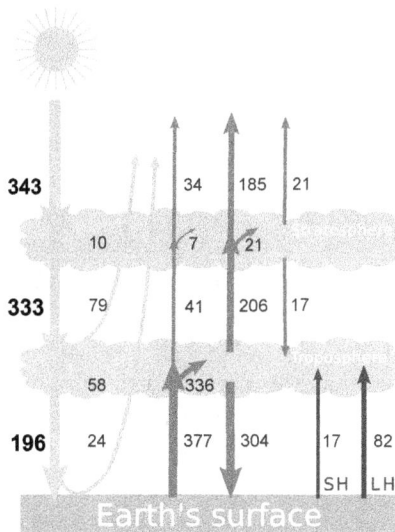

Figure 2.19   A schematic of the global energy balance. All values are in $\mathrm{W\,m^{-2}}$. Yellow arrows represent solar radiation, red arrows represent long λ terrestrial radiation and blue arrows represent fluxes of sensible (SH) and latent heat (LH).

Source: Adapted from J.M. Wallace and P.V. Hobbs, *Atmospheric Science: An Introductory Survey*, p. 420, Elsevier (2006).

particularly between tropical and polar latitudes, and over the course of a year. The latitudinal differences in albedo and insolation lead to heating imbalances, which drives atmospheric and oceanic circulation. Redistribution of energy across the Earth's surface occurs through the processes of *sensible heat (SH) flux, latent heat (LH) flux*, and *surface heat flux* into oceans. SH flux is the transfer of heat energy from the surface to the atmosphere by conduction and convection. This energy then moves from the tropics to the poles by bulk flow of air, creating atmospheric circulation. Latent heat flux transfers energy by converting ice and liquid water into vapour, which can be moved by atmospheric circulation both vertically and horizontally to colder locations. Here, it condenses as rain or is deposited as snow, thereby releasing the heat energy stored within it. Finally, large quantities of radiation are absorbed at the surface of the tropical oceans and converted into heat energy; water from the warmed surface is then transferred downward into the water column by conduction and convection, while horizontal transfer from the equator to the poles is mediated by ocean currents.

In addition to the transport of heat from one part of the planetary surface to another, there must also be transport of heat from the surface and lower atmosphere back to space; in order for the Earth's temperature to remain stable over long periods of time, the incoming and outgoing energy at the TOA have to be equal, i.e. the (climate) system must be in radiative equilibrium. Around $103\,W\,m^{-2}$ of the incoming radiation flux is reflected at the TOA back to space by a combination of clouds, atmospheric aerosol, air molecules, and bright surfaces such as sea ice and snow — this is the albedo. Around $240\,W\,m^{-2}$ of solar intensity is therefore absorbed by the Earth's climate system, with approximately $68\,W\,m^{-2}$ absorbed in the atmosphere by water vapour, clouds and ozone, while the remaining $172\,W\,m^{-2}$ is absorbed by the surface. Clearly, $240\,W\,m^{-2}$ must be emitted back to space in order that the surface temperature remains constant.

Measurements show that the Earth's surface radiates a net flux of only $34\,W\,m^{-2}$ back to space, while the atmosphere radiates a much higher amount of *ca.* $206\,W\,m^{-2}$. Comparing these values with the absorbed fluxes of 172 and $68\,W\,m^{-2}$ for the surface and atmosphere, respectively, we can conclude that solar heating mostly occurs at the surface, while radiative cooling in the form of IR emission mostly occurs in the atmosphere. It is therefore pertinent to consider how this redistribution of energy between the surface and the atmosphere occurs. Energy balance at the planetary surface requires that the land and ocean dispose of the energy absorbed in the incident $172\,W\,m^{-2}$ of solar radiation flux. The surface does this through a combination of evaporation, convection, and emission of IR radiation: evaporation of liquid water accounts for around $82\,\mathrm{W\,m^{-2}}$, while $17\,\mathrm{W\,m^{-2}}$ leaves through convection and the emission of thermal IR radiation removes the remaining $73\,\mathrm{W\,m^{-2}}$.

As we saw in Section 2.3.2, energy balance must occur within the atmosphere. Satellite measurements show that the atmosphere radiates $206\,\mathrm{W\,m^{-2}}$ of IR intensity and therefore must also be absorbing the same amount. As noted above, clouds, aerosols, water vapour, and ozone absorb $68\,\mathrm{W\,m^{-2}}$ while evaporation and convection transfer $99\,\mathrm{W\,m^{-2}}$ from the surface to the atmosphere, constituting a total of $167\,\mathrm{W\,m^{-2}}$. The remaining $39\,\mathrm{W\,m^{-2}}$ required to ensure radiative balance comes from the Earth's surface, which radiates the net equivalent of $73\,\mathrm{W\,m^{-2}}$ as IR. As noted above, around $34\,\mathrm{W\,m^{-2}}$ escapes directly to space, with the remaining $39\,\mathrm{W\,m^{-2}}$ being transferred to the atmosphere by GHG absorption of the IR radiation emitted by the surface. When GHGs absorb IR radiation, their temperature rises, and they radiate an increased amount of IR in *all* directions. Radiation emitted upwards encounters further GHG molecules that can absorb IR and then re-radiate, although by an altitude of *ca.* 5–6 km, their concentration is sufficiently small that they can generally radiate freely to space. Some of the IR radiation propagates downward and is ultimately absorbed by the surface, resulting in the surface temperature being warmer than it would be if it were heated only by direct solar heating — this is the natural greenhouse effect. We emphasise that observations show that the flux of downwelling IR radiation from the atmosphere can exceed that due to the direct solar flux.

2.4.2 Radiative Forcing and Feedbacks

Any changes to the Earth's climate system that affect how much energy enters or leaves the system alters its radiative balance, and can force the global surface temperatures to rise or fall (eventually) to a new equilibrium state. These influences are called *climate forcings*. Examples of natural climate forcings include changes in the Sun's brightness (known as *Milankovitch cycles*) and large volcanic eruptions that inject aerosol particles into the stratosphere. Anthropogenic forcings include particulate emissions (particles absorb and/or reflect incoming solar radiation), deforestation (which changes the reflectivity of the Earth's surface), and the increasing concentration of CO_2 and other GHGs (which decrease the amount of heat radiated to space). In addition, a forcing can trigger *feedbacks* that amplify or reduce the original forcing. For example, changes in the mixing ratio of IR or UV absorbers can alter the atmospheric temperature thereby altering circulation, transport and surface process such as emissions and deposition. Subsequent changes in the rates of chemical reactions then leads to further variation in atmospheric composition, and constitutes feedback.

Climate or radiative forcing (RF) is quantified by the net radiative flux imbalance, ΔF, at the top of the troposphere (although in some cases the imbalance is taken as being at the TOA) when an instantaneous perturbation is applied. When forcing occurs, for example as a result of increasing the mixing ratio of CO_2, the most important question is *by how much does this cause the surface temperature to change?* This question is extremely difficult to answer precisely, and lies at the heart of contemporary climate science. By convention, the temperature change, ΔT_g, is determined by the *equilibrium climate sensitivity*, S, defined as the global mean surface warming in response to a doubling of $[CO_2]$ after the system has reached a new steady state. It should be appreciated that climate sensitivity cannot be measured directly, but can be estimated from comprehensive climate models.

Within the simplest energy balance model, a forcing, ΔF, leads to heating, ΔQ, in the system, with $\Delta Q = \Delta F - \lambda' \Delta T_g$, where $\lambda' \Delta T_g$ represents the increased outgoing IR radiation induced by the forcing and which is assumed to be proportional to the amount of surface warming, ΔT_g. Heat is taken up largely by the oceans and for a constant forcing, the system eventually approaches a new equilibrium in which the heat uptake ΔQ is zero and the RF is balanced by additional emitted IR radiation. Calculations show that the forcing is related to the change in CO_2 abundance according to the relationship $\Delta F \approx 5.3 \ln(C/C_0)$, where C_0 and C are initial and final CO_2 mixing ratios.[11] Hence S, the equilibrium global average temperature change for a doubling of the CO_2 concentration (usually relative to pre-industrial levels), corresponds to a forcing of *ca.* $3.7 \, \mathrm{W \, m^{-2}}$. By convention, the ratio of the equilibrium temperature change to the forcing, $\Delta T / \Delta F = 1/\lambda'$, is known as the *climate*

---

[11] ΔF is only a slowly varying function of $[CO_2]$, since the CO_2 IR bands are very strongly absorbing and so only additional absorption in the wings of the transitions, which are not yet saturated, will contribute. Thus, although CO_2 has the highest overall RF, its RF *per molecule* is low (see Section 2.3.3). For comparison, the RF associated with CFCs varies linearly with concentration, a behaviour typical of weakly absorbing transitions occurring within the atmospheric window.

sensitivity parameter. Its value is highly model dependent, with the IPCC Fourth Assessment Report summarising climate sensitivity as *likely to be in the range 2–4.5°C with a best estimate of about 3°C, and is very unlikely to be less than 1.5°C.*

As noted at the start of this section, the changes in flux of outgoing IR radiation that balance the change in forcing are influenced by climate feedbacks and the sensitivity, S, can then be recast as $S = \Delta T_0/(1-f)$, where f is the *total feedback factor* and ΔT_0 is the calculated rise in temperature of a pseudo-black body atmosphere (with $\varepsilon \approx 0.5$) in response to a doubling of $[CO_2]$; circulation models estimate this rise as $1.2\,\text{K}$. Importantly, feedback can either amplify (if $0 < f < 1$) or damp the black body response. The feedback factor f can be approximated as the sum of all the individual feedbacks, f_i, which include increases in water vapour levels with warming, as well as changes in albedo, cloud cover, and lapse rate. A summary of the principal radiative forcing agents of climate change over the last 260 years is given in Figure 2.20. Particularly noteworthy are the sizes of the error bars associated with each component, which reflect the uncertainty in each of the respective values. To first-order, the feedbacks are temperature independent, leading to a climate sensitivity that is constant with time, and a global temperature response from different forcings that is approximately additive. However, we note that some feedbacks will change with temperature, which means that the assumption of a linear feedback term $(\lambda' \Delta T_g)$ is valid only for temperature changes of a few degrees.

The concept of radiative forcing, feedbacks and temperature response is illustrated in Figure 2.21. When a forcing such as increased GHG concentrations perturbs the energy balance, it does not change the value of T_g instantaneously — it can take years or even decades for the full impact of a forcing to be realised (see Figure 2.21). This lag between the time at which a perturbation occurs and the time when the impact on T_g becomes fully apparent is due to the large heat capacity of the oceans; the top $3\,\text{m}$ of the ocean has the same heat capacity as the entire atmosphere, and slow deep mixing delays the effect of any warming. It is estimated that only about 60% of the effect of industrial activity since 1700 is manifest in surface temperature increases to date. This means that even if CO_2 levels were stabilised, further increases in atmospheric temperature of around $1\,\text{K}$ are expected. The heating effect of CO_2 is compounded by the decreasing solubility of CO_2 in water with increasing temperature. Importantly, the oceans contain around 50 times more CO_2 than the atmosphere, in a variety of forms, primarily HCO_3^- (95%), CO_3^{2-} (3.3%), and dissolved CO_2 (1.7%).

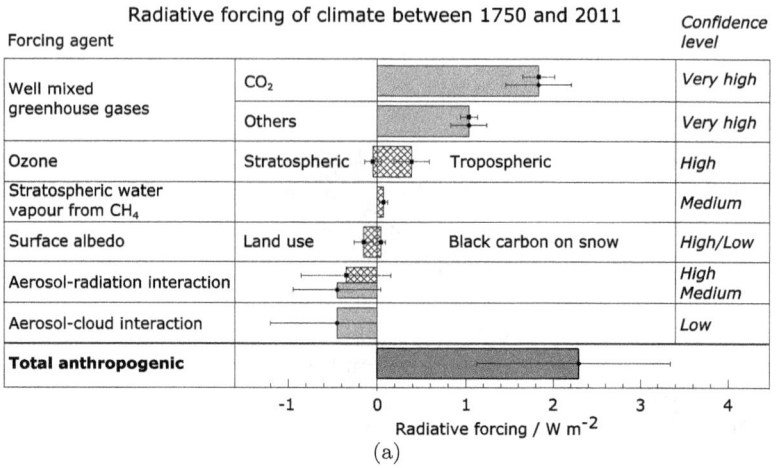

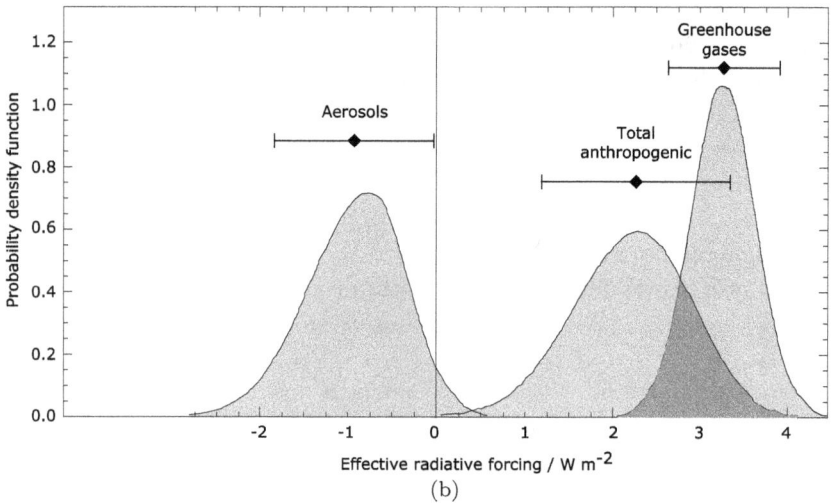

Figure 2.20   (a) Summary of the principal components of RF. (b) Probability density functions associated with the effective RFs attributable to aerosols and clouds, GHGs and anthropogenic activity; the effective forcing from aerosols and clouds partly offsets that from GHGs but has a higher uncertainty associated with it.

Source: Adapted from p. 697 of G. Myhre *et al. Anthropogenic and Natural Radiative Forcing. In*: *Climate Change 2013*: *The Physical Science Basis. Contribution of Working Group I to the Fifth Assessment Report of the Intergovernmental Panel on Climate Change* [Stocker, T.F., D. Qin, G.-K. Plattner, M. Tignor, S.K. Allen, J. Boschung, A. Nauels, Y. Xia, V. Bex and P.M. Midgley (eds.)]. Cambridge University Press, Cambridge, United Kingdom and New York, NY, USA.

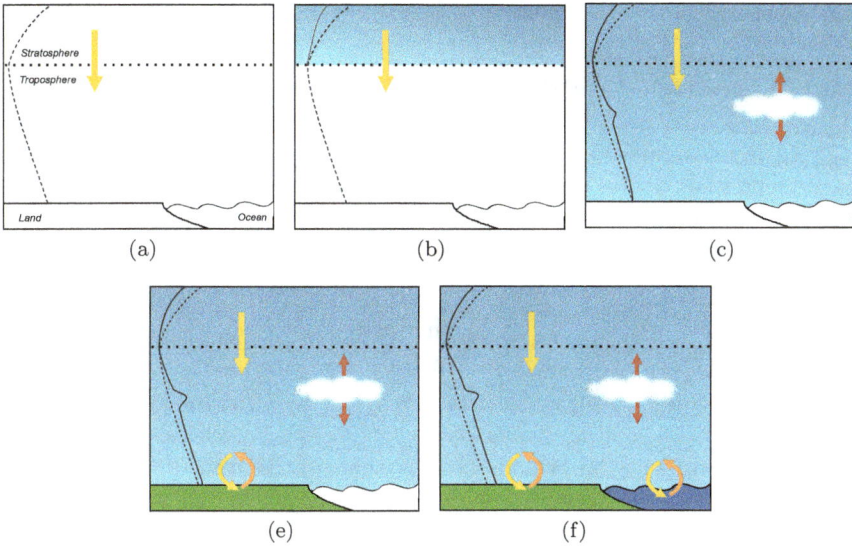

Figure 2.21 A schematic of the timeline for RF. (a) RF begins with a radiation flux imbalance (yellow colour arrow) at the tropopause. The unperturbed temperature profile is shown as a dashed line; (b) the stratosphere adjusts relatively quickly (days to months) to the forcing and there is a change in the stratospheric lapse rate as denoted by the solid grey line; (c) the troposphere then responds to the forcing (days to months) and there is a change in both heating rate and cloud formation which results in a change in the atmospheric temperature profile; (d) and (e) the surface temperature increases and the Earth emits increased amounts of long λ radiation while large amounts of heat is taken up by the oceans; these processes occur on the timescale of decades. Equilibrium, i.e. no tropospheric flux imbalance, is (eventually) reached on the timescale of centuries.

Source: Adapted from R. Knutti and G. Hegerl, *Nature Geoscience* **1** 735 (2008).

There is a clear need to reduce the amount of CO_2 and other GHG emissions into the atmosphere, and considerable effort continues to be expended in trying to predict the response of the climate to past, present and future emissions. Many climate change models exist for a range of different CO_2 emission scenarios, and all predict an increase in global surface temperature of between 1.5°C and 6°C by 2100. The largest temperature changes are expected at high latitudes in the Northern hemisphere, as a result of changes in snow cover. Temperatures here are around 0°C and so a slight temperature increase leads to significant reduction in snow cover, thereby reducing the albedo in another important example of positive feedback.

2.5 Questions

2.5.1 Essay-style Questions

Q2.1: How is a photolysis rate defined? Hence state an equation for the lifetime of a molecule undergoing photolysis. Why is N_2O only photolysed in the stratosphere while NO_2 is photolysed in both the stratosphere and troposphere?

Q2.2: Explain what is meant by the term *black body radiator*. Why are good emitters of thermal radiation also good absorbers of thermal radiation?

Q2.3: Explain why the principal components of the atmosphere (nitrogen and oxygen) do not absorb mid-IR radiation whereas water vapour and carbon dioxide are strong absorbers.

Q2.4: The carbon dioxide level in the atmosphere has risen from 280 ppmv before industrialisation to a current level of 400 ppmv. Explain why this could lead to an increase in the Earth's mean surface temperature.

Q2.5: The mean temperature of the Earth's surface is higher than the mean temperature expected from the balance of absorbed and emitted radiation. Explain qualitatively why this is so and identify the atmospheric constituents which make major contributions to surface warming. Discuss whether or not the contributions of different gases are additive, and how the warming effect may depend on their concentrations.

Q2.6: Explain what is meant by the term *feedback mechanism* in the context of atmospheric science. Illustrate your answer by describing one positive and one negative feedback mechanism, explaining briefly in each case how the mechanism operates.

Q2.7: The gravitational acceleration on Venus is roughly the same as on Earth and its effective emitting temperature is also similar. Its atmosphere is mainly CO_2 which has approximately the same specific heat capacity as air. The surface temperature on Venus is observed to be 735 K and the surface atmospheric pressure is 92 bar. Explain why the stratosphere on Venus begins at an altitude which is 6 times higher than that on Earth.

Q2.8: Calculations suggest that a doubling of the atmospheric CO_2 concentration will lead to a decrease in temperature in the stratopause of about 10 K. Explain schematically why the temperature might decrease in the stratosphere.

Q2.9: An IR radiometer observing the Earth finds that the upwelling radiance is a maximum for a wavelength of *ca.* $10\,\mu m$. How may this observation be interpreted in terms of the temperature structure of the atmosphere?

2.5.2 Problems

P2.1: The most intense feature in the Earth's dayglow is emission from the transition $O_2(^1\Delta_g) \to O_2(^3\Sigma_g^-)$.

(a) How is the excited molecular oxygen $O_2(^1\Delta_g)$ formed?

(b) The $\nu' = 0 \to \nu'' = 0$ vibrational band of this system is the most intense as observed from space, but very weak as seen from the ground, whereas the $\nu' = 0 \to \nu'' = 1$ band can be observed equally well from outside or within the atmosphere. Suggest a reason.

(c) The IR atmospheric band originates at altitudes around 50 km, where the gas density is 2×10^{16} molecules cm^{-3}, from the transition $O_2(^1\Delta_g) \to O_2(^3\Sigma_g^-)$ whose radiative lifetime is 44 minutes. In a total solar eclipse, the intensity of the band was found to drop with a half-life of 108 s. Obtain a rate constant for quenching of the excited state and comment on the result. [*Note:* A typical quenching rate constant is 3×10^{-11} cm^3 molecule^{-1} s^{-1}.]

P2.2: A volcanic eruption produced an aerosol layer in the stratosphere that was 1 km thick and had an optical depth of 0.2 at the wavelength 500 nm. Assuming that the particles are $1\,\mu m$ in diameter, calculate the size parameter of the particles, and hence estimate their number density within the volcanic aerosol layer.

P2.3: (a) Figure 2.3 shows the UV absorption spectrum of ozone. Assuming an initial ozone column of 7.8×10^{18} molecule cm^{-2}, calculate the percentage increase in light intensity reaching the Earth's surface at 300 nm for a 20% decrease in the ozone column. Assume a solar zenith angle of $45°$.

(b) Figure 2.3 also shows the absorption spectrum for O_2. Use both spectra to compare the light attenuation due to absorbance by O_2 and O_3 with Rayleigh scattering for wavelengths in the range 150–350 nm. Assume an altitude of $z = 0$, a solar zenith angle of $0°$, and an ozone column of 7.8×10^{18} molecule cm^{-2}.

P2.4: The spectral radiance for a black body at temperature T is given by:

$$B_\nu(T) = \left(\frac{2h\nu^3}{c^2}\right)\left(\frac{1}{\exp\left(\frac{h\nu}{k_BT}\right) - 1}\right)$$

Integrate the spectral radiance over the entire frequency range using the substitution $x = \frac{h\nu}{k_BT}$ and comment on the result. You will require the following integral:

$$\int_0^\infty \frac{x^3 dx}{e^x - 1} = \frac{\pi^4}{15}$$

P2.5: (a) Assume that the surface of the Sun acts as a black body radiator with a temperature of 5780 K.

(i) The distance from the Earth to the Sun is 1.5×10^{11} m, the solar radius is 7.0×10^8 m and the Stefan–Boltzmann constant, σ, is 5.67×10^{-8} W m^{-2} K^{-4}. Use these data to show that the solar energy flux at the Earth's orbit is 1376 W m^{-2}.

(ii) Explain why the average flux at the surface of the Earth is 241 W m^{-2}.

(b) Estimate the temperature at the Earth's surface if the Earth behaves as a perfect black body with an atmosphere consisting of only N_2 and O_2.

(c) Estimate how much the Earth's mean surface temperature would change if

(i) the radius of the Earth's orbit about the Sun was to increase by 10%, and

(ii) the Earth's albedo were to reduce to 0.25.

(d) Suppose that the Sun stops shining. Neglecting the heat from the earth's core, show that the rate of loss of temperature T of the Earth with time t will be

$$\frac{dT}{dt} = \frac{\sigma A T^4}{MC}$$

where A and M are the surface area and mass of the Earth, and C is the average specific heat capacity of the material of the Earth. Show that the time taken for the effective temperature to fall by a factor of 2 is given by

$$t = \frac{7MC}{3\sigma A T^3}$$

P2.6: Consider a one-layer model of the atmosphere with transmission τ_{sol} at solar wavelengths and transmission τ_{IR} at terrestrial wavelengths. Stating any assumptions you make, show that the surface temperature T_g can be estimated from

$$T_g = \left[\left(\frac{0.71 F_{\text{in}}}{\sigma} \right) \left(\frac{1 + \tau_{\text{sol}}}{1 + \tau_{\text{IR}}} \right) \right]^{1/4}$$

where F_{in} is the insolation and σ is the Stefan–Boltzmann constant. Calculate the surface temperature given that $F_{\text{in}} = 343\,\text{W}\,\text{m}^{-2}$, $\tau_{\text{sol}} = 0.9$, and $\tau_{\text{IR}} = 0.2$.

P2.7: (a) Outline briefly the mechanisms of Doppler and pressure broadening.

(b) The mass-weighted absorption coefficient for Doppler broadening is defined as

$$k(\nu) = \left(\frac{s}{\alpha_D \sqrt{\pi}} \right) \exp\left(-\left(\frac{\nu - \nu_0}{\alpha_D} \right)^2 \right)$$

where

$$\alpha_D = \left(\frac{\nu_0}{c} \right) \left(\frac{2 k_B T}{M} \right)^{1/2}$$

For comparison, the mass-weighted absorption coefficient for pressure broadening is defined as:

$$k(\nu) = \left(\frac{s}{\pi} \right) \frac{\gamma}{(\nu - \nu_0)^2 + \gamma^2}$$

(i) Define the variables used in the equations above.

(ii) What is meant by the *equivalent width* of a spectral line?

(iii) Show that the equivalent width of a pressure-broadened line is proportional to m and $\sqrt{m}$ in the *weak* and *strong* limits, respectively, where m is the amount of absorber in units of mass per unit area.

You will require the following integral:

$$\int_{-\infty}^{\infty} \left(1 - e^{ax^2}\right) x^{-2} \mathrm{d}x = 2\sqrt{\pi a}$$

where $x = (\nu - \nu_0)/\gamma$ and $a = mS/\gamma\pi$

(c) A CO_2 transition at $667\,\mathrm{cm}^{-1}$, has a halfwidth at half maximum at standard temperature and pressure (STP), α_0, of $0.1\,\mathrm{cm}^{-1}$. Find the approximate pressure level in the atmosphere at which the transmittance over a horizontal path at the line centre under conditions of pure Doppler broadening is the same as that under pure pressure broadening by a line of the same strength. How would you expect the transmission due to Doppler and pressure broadening to compare as a function of altitude? Assume that the atmospheric temperature is $255\,\mathrm{K}$ and the halfwidth, γ, depends upon pressure p and temperature T as follows:

$$\gamma = \alpha_0 \left(\frac{p}{p_0}\right) \left(\frac{T_0}{T}\right)^{1/2}$$

where α_0 is the halfwidth at standard temperature, T_0, and pressure, p_0.

Chapter 3

Stratospheric Chemistry

Introduction

The chemistry of the Earth's stratosphere was introduced briefly in Chapter 1. We will now focus on this upper region of the Earth's atmosphere and study stratospheric processes in detail. As noted in Section 2.1.3, molecular oxygen and ozone present in the stratosphere absorb strongly in the UV, shielding the lower reaches of the atmosphere and the Earth's surface from damaging radiation at these wavelengths. Absorption of UV light by these species yields highly reactive atomic oxygen, leading to a rich variety of radical chemistry. One consequence of this chemistry is formation of the ozone layer, within which the ozone concentration is at least an order of magnitude higher than the concentration at the tropopause.

The chemistry occurring in the stratosphere is dominated by processes involving O and O_3 (collectively known as O_x) and their reactions with hydrogen oxide radicals (HO_x), nitrogen oxides (NO_x), and Cl-containing species such as chlorofluorocarbons (CFCs), provide efficient gas-phase catalytic cycles which contribute to ozone destruction. CFCs are anthropogenic in origin and their presence has had far-reaching consequences, particularly being intimately involved in the formation of the ozone hole. The latter part of the chapter details the combination of heterogeneous chemistry and transport processes in the presence of Cl that contributes to this annual massive ozone deficit over Antarctica.

Table 3.1   Reactions in the Chapman cycle used for modelling the formation of ozone in the stratosphere.

Reaction	Rate constant	Relative rate
(1) $O_2 + h\nu \rightarrow O + O$	$J_1 \sim 5 \times 10^{-10}$ s^{-1} (at $z = 40$ km)	Slow
(2) $O + O_2 + M \rightarrow O_3 + M$	$k_2 = 6 \times 10^{-34}(300/T)^2$ cm^6 s^{-1}	Fast
(3) $O_3 + h\nu \rightarrow O + O_2$	$J_3 \sim 2 \times 10^{-3}$ s^{-1} (at $z = 40$ km)	Fast
(4) $O + O_3 \rightarrow 2O_2$	$k_4 = 1 \times 10^{-11} \exp(-2100/T)$ cm^3 s^{-1}	Slow

3.1 The Chapman Cycle: O_x on Its Own

3.1.1 Odd Oxygen

The simplest photochemical mechanism used for modelling the formation of the stratospheric ozone layer is the Chapman cycle. The constituent steps of the cycle are displayed in Table 3.1. The first step requires high-energy ($\lambda \leq$ 240 nm) photons to initiate photolysis of O_2 and, as discussed in Chapter 2, these wavelengths are only found in the higher reaches of the atmosphere. The cycle also relies on photolysis of O_3 (step (3) in Table 3.1), which occurs predominantly via a spin-allowed pathway to produce electronically-excited atomic oxygen, $O(^1D)$, and molecular oxygen, $O_2(a^1\Delta_g)$. Though the quantum yield for this pathway is *ca.* 0.9 across the 240–310 nm spectral range, the electronic quenching of $O(^1D)$ by N_2 and/or O_2 to form ground state $O(^3P)$ atoms is very rapid, and as such no differentiation is made between $O(^1D)$ and $O(^3P)$ in the Chapman reaction scheme. The photolysis rate constants J_1 and J_3, for photolysis of O_2 and O_3, respectively, vary with altitude as shown in Figure 3.1(a). Their calculation will be discussed further in Section 3.1.2.

The fastest steps in the Chapman cycle are steps (2) and (3), which interconvert O and O_3. These two reactions are said to preserve *odd oxygen* ($O_x = O + O_3$), and their relative rates determine the relative amounts of O and O_3 in the stratosphere. In Section 1.6.3, it was shown that a system of two interconverting species, A $\rightleftharpoons$ B, approaches equilibrium with time constant $\tau = 1/(k_1 + k_2)$, where k_1 and k_2 are the rate constants for the forward and backward processes respectively. For reaction (2) the rate is $k_2[O][O_2][M]$, with pseudo-first-order rate constant $k_2[O_2][M]$, while reaction (3) has rate $J_3[O_3]$ and first-order rate constant J_3. The time constant for the interconversion is therefore

$$\tau_{2,3} = \frac{1}{k_2[O_2][M] + J_3} \tag{3.1}$$

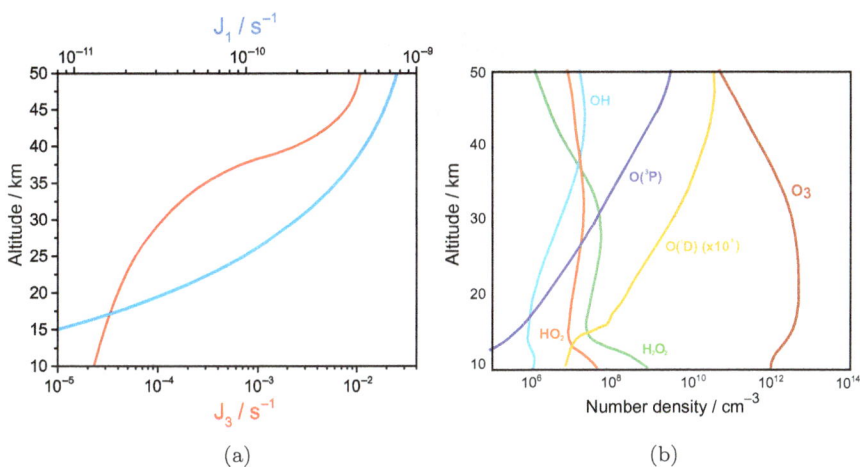

Figure 3.1 (a) Rate constants for the photolysis steps of the Chapman cycle and (b) the number densities of O_x as a function of altitude. The number densities for HO_x are also shown for comparison.

Source: Data taken from *Chemical Kinetics and Photochemical Data for Use in Stratospheric Modeling* by W.B. DeMore *et al.*, JPL Publication, 97–104 (1997).

For typical values of the rate constants at 40 km altitude, $\tau_{2,3}$ is *ca.* 0.5 s, so equilibrium is rapidly attained. The short induction time for achievement of equilibrium means that the steady-state approximation (SSA) (see Section 1.6.3) can always be applied to these two reactions, giving

$$\frac{d[O_3]}{dt} = k_2[O][O_2][M] - J_3[O_3] = 0 \qquad (3.2)$$

Therefore

$$\frac{[O]}{[O_3]} = \frac{J_3}{k_2[O_2][M]} \qquad (3.3)$$

and this relationship defines the ratio $[O]/[O_3]$ at all altitudes within the atmosphere. The number densities of O and O_3 are plotted as a function of altitude in Figure 3.1(b) and it is clear that ozone is present at a much higher concentration than atomic oxygen in both the troposphere and the stratosphere. The maximum $[O_3]$ is *ca.* 5×10^{12} molecule cm^{-3} and corresponds to a mixing ratio of ~ 10 ppmv (at 30 km altitude). Also, since the atomic oxygen concentration, $[O]$, is proportional to $1/([O_2][M])$, and therefore proportional to (pressure)$^{-2}$, the $[O]$ decreases very rapidly with

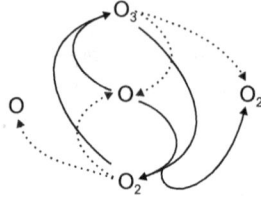

Figure 3.2   The Chapman cycle (dotted arrows indicate photolysis reactions).

decreasing altitude (increasing pressure). Reaction cycles involving atomic oxygen are therefore inefficient at low altitude. At night, [O] drops to zero, as there is no photolysis.

Since reactions (2) and (3) conserve odd oxygen (O and O_3), the ozone concentration is determined by the amount of odd oxygen available, which in turn depends on its lifetime, τ_{O_x}. Only the slow reactions (1) and (4) lead to a change in odd oxygen concentration; photolysis of O_2 in reaction (1) leads to formation of two odd oxygen species, O + O, while reaction (4) removes two members of the odd oxygen family, O + O_3 (see Figure 3.2). Thus, the rate of change of odd oxygen concentration is

$$\frac{d([O] + [O_3])}{dt} = \frac{d[O_x]}{dt} = 2J_1[O_2] - 2k_4[O][O_3] \tag{3.4}$$

The timescale for attaining a steady-state concentration of odd oxygen varies with altitude: in the upper stratosphere ($>\sim 60\,\text{km}$), equilibration occurs within hours, and therefore use of the SSA is appropriate. In the lower stratosphere ($\sim 20\,\text{km}$), however, equilibration takes years, and a steady state is never established, as external factors such as temperature, transport and intensity of solar radiation all vary much more quickly. Once a steady state has been reached in the middle and upper atmosphere, the rate of change of odd oxygen concentration can be set to zero in Equation (3.4). Using the atomic oxygen concentration from Equation (3.3), we obtain a steady-state ozone concentration given by

$$[O_3] = \left(\frac{J_1 k_2}{J_3 k_4}\right)^{1/2} [O_2][M]^{1/2} \tag{3.5}$$

Since $[O_3] \gg [O]$, we often make the approximation that $d[O_x]/dt \approx d[O_3]/dt$. While the Chapman model qualitatively reproduces the general shape of the vertical profile of O_3, as shown in Figure 3.3, it overestimates the steady-state ozone concentration by approximately a factor of 2 at

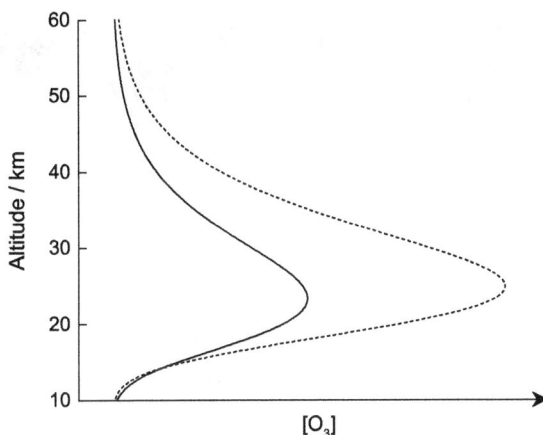

Figure 3.3 Schematic of stratospheric ozone abundance predicted by the Chapman reactions (dashed lines) compared to the actual profile.

all altitudes. Given that the input parameters to the calculation are well known, consisting of absorption cross-sections and reaction rate constants that can be measured accurately in laboratory studies, it must be concluded that the Chapman cycle does not account for all of the naturally occurring sinks for O_x. In fact, there are a number of additional reactions that remove O_x, involving species such as HO_x, NO_x, and ClO_x. These catalytic reactions will be discussed in detail in Section 3.2. Further shortcomings of the Chapman cycle are revealed in the variation of $[O_3]$ with latitude. At high latitudes, the rate constant J_1 for the initiation step in the Chapman cycle is small, and $[O_3]$ should be correspondingly low. However, even during the polar night, when J_1 is zero, the column densities of ozone are in fact found to be highest at these latitudes. In addition, while $[O_3]$ is over-predicted at tropical latitudes, on moving to extra-tropical latitudes it is under-predicted in the lower stratosphere but over-predicted in the upper stratosphere.

These observations may all be rationalised by the fact that the photo-chemical lifetime of O_3 is long enough (at altitudes below 30 km) for ozone to be transported by general circulation.[1] After entering the stratosphere

---

[1] As discussed in Chapter 1, tropospheric air enters the stratosphere at low latitudes. This is inferred from the fact that the stratosphere is very dry, and so air must have entered from the tropical tropopause at an altitude where the lowest tropospheric temperatures occur and water is frozen out.

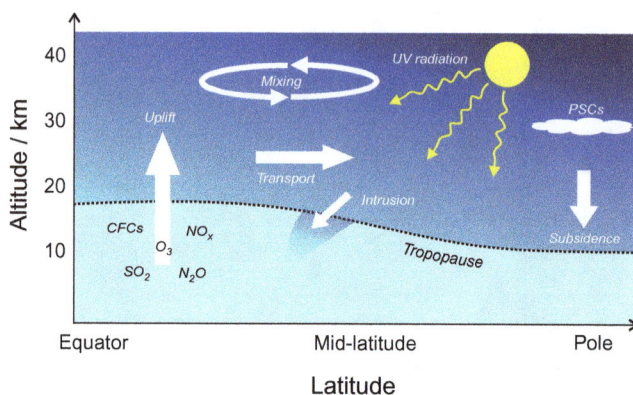

Figure 3.4   A schematic of ozone transport.

at low latitudes, the air moves polewards in the lower stratosphere and then descends back into the troposphere at middle and high latitudes as depicted in Figure 3.4. Poleward transport in the lower stratosphere is observable in the ozone distributions. In summary, although ozone is mainly formed in the equatorial stratosphere, significant number densities are found at high latitudes, indicating transport from low latitudes.

3.1.2 The Photolysis of O_2 and the Chapman Layer

The first step in the Chapman cycle is the UV photolysis of O_2, and we will now consider this important process in more detail. In particular, we will use the equations presented in Chapter 2 to derive the altitude dependence of the photolysis rate.

Consider a horizontal layer of atmosphere, at altitude z and of thickness dz, and assume that the Sun is directly overhead. The intensity of solar radiation at frequency ν reaching the layer, assuming no scattering losses, is given by

$$I_\nu(z) = I_{\nu,\infty} \exp\left(-\chi_\nu(z)\right) \tag{3.6}$$

The quantity $\chi_\nu(z)$ is the optical depth at altitude z, and is defined as

$$\chi_\nu(z) = \int_z^\infty k_\nu(z)\rho(z)\mathrm{d}z \tag{3.7}$$

where $\rho(z)$ is the gas density and $k_\nu(z)$ is the *mass-weighted* absorption coefficient (see Section 2.2.1). From the Beer–Lambert law, the absorption rate within the layer is simply

$$\frac{dI_\nu(z)}{dz} = I_\nu(z)k_\nu(z)\rho(z) \tag{3.8}$$

If k_ν is independent of z, and $\rho(z)$ is well described by the hydrostatic equation, i.e. $\rho(z) = \rho_0 \exp(-z/H)$, with H the scale height, then

$$\chi_\nu(z) = Hk_\nu\rho_0 \exp(-z/H)$$
$$= \chi_\nu(0) \exp(-z/H) \tag{3.9}$$

This result shows that the optical depth increases as the solar radiation penetrates downwards, in line with the discussion in Chapter 2. Substituting Equation (3.9) into Equation (3.8) yields a new expression for the rate of absorption

$$dI_\nu(z)/dz = I_{\nu,\infty}k_\nu(z)\rho_0 \exp(-[(z/H) + Hk_\nu\rho_0 \exp(-z/H)]) \tag{3.10}$$

Assuming a quantum yield for photolysis of unity, the rate of photodissociation is simply equal to the rate of absorption.

Figure 3.5 shows the variation of the optical depth χ_ν, the actinic light intensity I_ν, and the photolysis rate as functions of altitude. We see that the photolysis rate peaks at a well-defined altitude. We can determine this altitude by maximising Equation (3.10) with respect to z, and find that the maximum photolysis rate occurs at the altitude for which the optical depth is unity. This region of the atmosphere is known as the *Chapman layer*. The altitude dependence of the photolysis rate as we move vertically through the Chapman layer results from the interplay of two factors. At high altitudes, there is a large actinic flux but few O_2 molecules, while at low altitudes the reverse is true. In each case, the photolysis rate is small. However, near the level of unit optical depth, both the flux and the absorber density are significant, and the photolysis rate reaches a maximum. Finally, we note that as well as describing the rate of photon absorption by O_2 as a function of altitude, Equation (3.10) also forms the basis of calculations to determine the UV heating rate of the stratosphere.

3.1.3 Measuring the O_3 Distribution: Remote Sounding

In order to evaluate the success of any chemical model of the stratosphere, the predictions of the model must be compared with experimental data.

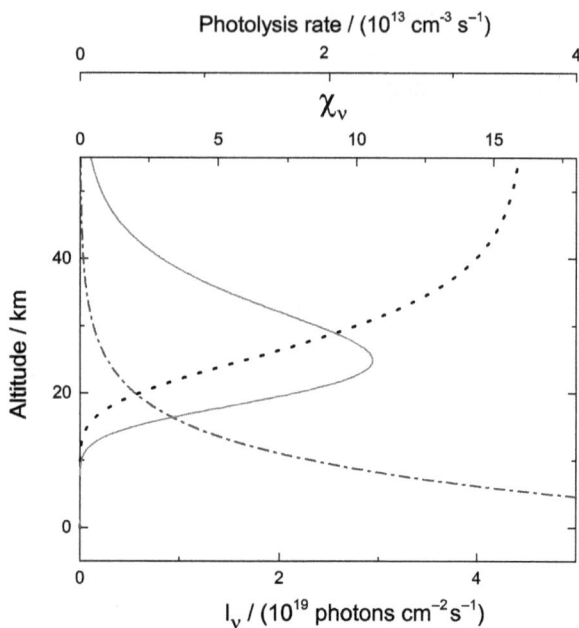

Figure 3.5   The optical depth χ_ν (dot-dashed line), the intensity (actinic) I_ν (dashed line), and the photolysis rate (solid line) as a function of altitude.

The distribution of ozone within the stratosphere is most commonly measured using a *Dobson spectrophotometer*.[2] O_3 absorbs strongly at wavelengths < 310 nm within the Hartley band. These wavelengths are completely absorbed within the stratosphere, and therefore cannot be used in any Earth-based measurement of stratospheric ozone concentration. However, ozone also absorbs weakly around 320 nm within the Huggins band and an appreciable amount of solar radiation at these wavelengths does reach the Earth's surface. Dobson realised that if the absorption cross-section, $\sigma(\lambda)$, of O_3 in the Huggins band is known, then by measuring the attenuation of these wavelengths by the atmosphere, the total number of O_3 molecules between the measuring instrument and the top of the atmosphere (TOA) may be determined, as we will show below. The resulting *column amount* is widely reported in *Dobson Units* (DUs) as defined in Section 1.2.1, and typical values are *ca.* 300 DU.

---

[2]See, for example, G.M.B. Dobson, *Appl. Opt.* **7** 387 (1968).

The attenuation of radiation at wavelength λ_1 over an absorbing path is simply $\exp(-\sigma(\lambda_1)U)$, where U is the column amount. The photon flux at the surface, $I_{\lambda_1,0}$, is therefore

$$I_{\lambda_1,0} = I_{\lambda_1,\infty} \exp\left(-\sigma\lambda_1 U\right) \tag{3.11}$$

where $I_{\lambda_1,\infty}$ is the actinic flux at the TOA. In practice, rather than making measurements at a single wavelength, measurements are made at two nearby wavelengths, typically separated by around 20 nm, and a ratio is taken, i.e.

$$\frac{I_{\lambda_1,0}}{I_{\lambda_2,0}} = \frac{I_{\lambda_1,\infty}}{I_{\lambda_2,\infty}} \exp(-(\sigma(\lambda_1) - \sigma(\lambda_2))U) \tag{3.12}$$

The ratio of incoming solar intensities, $(I_{\lambda_1,\infty}/I_{\lambda_2,\infty})$, is constant and can be measured, as is the ozone absorption cross-section $\sigma(\lambda)$. By measuring the ratio of light intensities $I_{\lambda_1,0}/I_{\lambda_2,0}$ reaching the Earth-based detector, the amount of ozone, U, in the (slanted) line of sight between the spectrometer and the Sun can be determined. The column amount in the vertical column of ozone is then $U/\cos\theta$, where θ is the solar zenith angle.

When making measurements on ozone, wavelengths of 305 and 325 nm are often used, as the absorption cross-sections differ by an order of magnitude at these two wavelengths. Using the difference between two wavelengths minimises potential errors in the measured column densities due to the presence of any additional absorbing species such as aerosols whose cross-sections do not vary as markedly with wavelength and altitude as that for O_3. As discussed in Chapter 2, incoming radiation is also attenuated due to Rayleigh scattering, as well as scattering and absorption by cloud particles. The data analysis can be modified to account for these effects. The now routine ground-based measurements of ozone column amounts are confirmed by data from satellite observations with the total ozone mapping spectrometer (TOMS) (see Figure 3.17 in Section 3.3).

3.2 Stratospheric Ozone Loss: HO_x, NO_x, and ClO_x Gas-phase Chemistry

Having considered simple models based on oxygen-only chemistry of the stratosphere, we now move on to consider additional O_x removing reactions involving HO_x, NO_x, and ClO_x radical species. These reactions are *catalytic*, and are therefore extremely efficient, despite the relatively low number densities of the species involved. We will highlight the *sources*, *sinks*, and *reservoirs* of these catalytic compounds, as well as the efficiency with which they react, which varies with altitude within the stratosphere.

3.2.1 Oxides of Nitrogen, NO_x

We begin by considering the role of NO and NO_2, collectively known as NO_x. Typical NO_x mixing ratios are around 10 ppbv, significantly less than typical O_3 ratios of up to 10 ppmv. NO and NO_2 are readily interconverted — see below — on a timescale of $\sim$100 s in sunlight, and so can be well approximated as being in steady state. NO is present in the stratosphere primarily due to oxidation of N_2O by the reaction $N_2O + O(^1D) \to 2NO$. Nitrous oxide, N_2O, is produced at the Earth's surface by bacteria and from fertilisers, and is well mixed in the troposphere due to its long photochemical lifetime ($\sim$100 years) and insolubility. Both factors mean that natural and anthropogenic emissions of N_2O are efficiently transported into the stratosphere. Once there, N_2O is either photolysed to produce $N_2 + O(^1D)$ or reacts with $O(^1D)$ to form NO. Once formed, NO can react with O_3 to make NO_2, which can subsequently react with $O(^3P)$ atoms (generated by quenching of $O(^1D)$ formed from O_3 photolysis) to regenerate NO in a catalytic cycle:

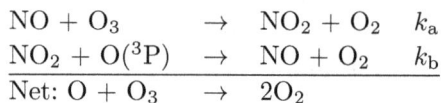

$$
\begin{array}{llll}
NO + O_3 & \to & NO_2 + O_2 & k_a \\
NO_2 + O(^3P) & \to & NO + O_2 & k_b \\
\hline
\text{Net: } O + O_3 & \to & 2O_2 &
\end{array}
$$

This cycle constitutes a loss route for O_x. The second step of the cycle is the rate-limiting step, as this occurs in competition with another set of reactions that interconvert NO_2 and NO, initiated by photolysis of NO_2 at wavelengths below 410 nm:

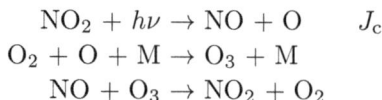

$$
\begin{array}{ll}
NO_2 + h\nu \to NO + O & J_c \\
O_2 + O + M \to O_3 + M & \\
NO + O_3 \to NO_2 + O_2 &
\end{array}
$$

This reaction cycle constitutes a *null reaction* overall, with all reactants regenerated as products. However, the competition between this reaction and the catalytic destruction of ozone by NO becomes important at lower altitudes. We will return to a discussion of the rapid cycling between NO and NO_2 in Section 4.4.

Taking both reaction schemes into account, the rate of NO loss is given by

$$\frac{d[NO]}{dt} = -k_a[NO][O_3] + k_b[NO_2][O] + J_c[NO_2] \qquad (3.13)$$

An equilibrium is rapidly reached between NO and NO_2, such that we can set $d[NO]/dt = 0$, yielding

$$k_a[NO][O_3] = k_b[NO_2][O] + J_c[NO_2] \qquad (3.14)$$

The rate of odd oxygen destruction is

$$\frac{d[O_x]}{dt} = -k_a[NO][O_3] - k_b[NO_2][O] + J_c[NO_2] \qquad (3.15)$$

Comparing Equations (3.13) and (3.15) then yields

$$\frac{d[O_x]}{dt} = -2k_b[NO_2][O] \qquad (3.16)$$

This analysis shows that odd oxygen loss occurs at twice the rate of the rate-determining step, with the rate of this step determined by the competition between the two reaction cycles described above. $[O]/[O_3]$ is still determined by the Chapman cycle.

The catalytic NO_x cycle is *terminated* by loss of NO_x in a three-body reaction of NO_2 with OH to produce nitric acid:

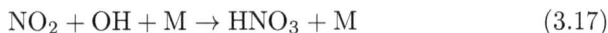

$$NO_2 + OH + M \rightarrow HNO_3 + M \qquad (3.17)$$

The HNO_3 product then acts as a reservoir for NO_x, holding the reactive, ozone-destroying NO_x radicals in an unreactive form. Other notable reservoirs are N_2O_5 and $ClONO_2$. The former is formed via the following pair of reactions:

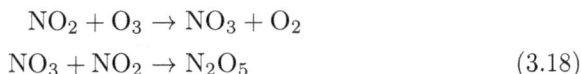

$$NO_2 + O_3 \rightarrow NO_3 + O_2$$
$$NO_3 + NO_2 \rightarrow N_2O_5 \qquad (3.18)$$

Formation of N_2O_5 by this route can only occur at night, as NO_3 is rapidly photolysed during the daytime to form $NO_2 + O(^3P)$ or $NO + O_2$, thereby regenerating NO_x (see Figure 3.6).

HNO_3 and N_2O_5 are relatively stable species (though only at night-time in the case of N_2O_5), with longer lifetimes than their NO_x precursors, but are eventually converted back to NO_2 by photolysis. Photolysis of HNO_3 at wavelengths below 310 nm yields NO_2 and OH products, while photolysis of

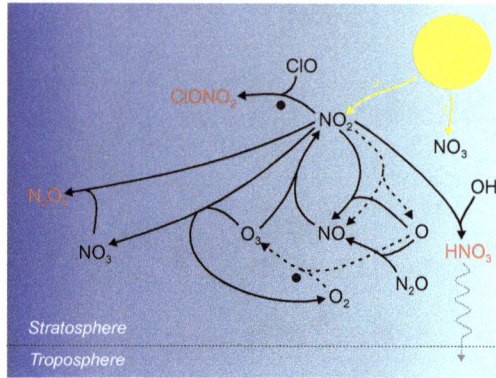

Figure 3.6   A schematic of the NO_x cycle.

Note: Reservoir species are in red; dotted lines represent key reactions of the null cycle.

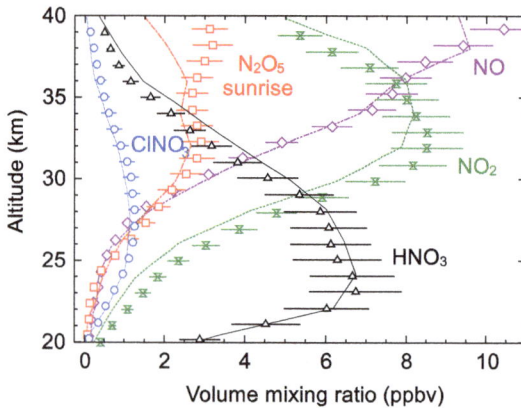

Figure 3.7   Volume mixing ratios of NO_x reservoirs and their altitude dependence.

Source: From World Meteorological Organisation (WMO) report (1998), Chapter 6 http://www.esrl.noaa.gov/csd/assessments/ozone/1998/chapters/chapter6.pdf adapted from B. Sen *et al.*, *J. Geophys. Res.* **103** 3571 (1998).

N_2O_5 yields NO_3 and NO_2 products. HNO_3 can also undergo oxidation by OH radicals to yield NO_3 and H_2O, followed by photolysis of NO_3. HNO_3 and N_2O_5 therefore act as temporary reservoirs of NO_x in the stratosphere. Permanent removal of NO_x, HNO_3, and N_2O_5 (collectively known as NO_y) from the stratosphere is achieved by the slow transport of these species to the troposphere, where deposition occurs. The important nitrogen reservoir species in stratospheric NO_x chemistry are shown in Figure 3.7, along with their abundance as a function of altitude.

As noted previously, N_2O has a long atmospheric lifetime. At present, N_2O mixing ratios are increasing by approximately 0.5% per year, thereby increasing stratospheric NO_x as well as acting as a very effective greenhouse gas. As will be discussed in Chapter 4, industrialisation leads to copious emission of NO_x into the troposphere. However, little of this reaches the stratosphere, with most of it being converted to HNO_3, which is then rained out.

3.2.2 Oxides of Hydrogen, HO_x

While the majority of the $O(^1D)$ product from photolysis of ozone is collisionally quenched to produce $O(^3P)$ atoms, a small fraction reacts with water vapour, present at relatively low mixing ratios (1–5 ppmv) to produce OH radicals:

$$O(^1D) + H_2O \rightarrow 2OH \qquad (3.19)$$

These radicals can then react catalytically with odd oxygen in a reaction cycle analogous to that for the NO_x radicals, i.e.

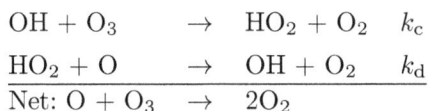

$$
\begin{array}{lll}
OH + O_3 & \rightarrow & HO_2 + O_2 \quad k_c \\
HO_2 + O & \rightarrow & OH + O_2 \quad k_d \\
\hline
\text{Net: } O + O_3 & \rightarrow & 2O_2
\end{array}
$$

As was the case for the NO/NO_2 cycle, the OH/HO_2 cycle removes two odd oxygen species per cycle, as shown in Figure 3.8, and regenerates OH.

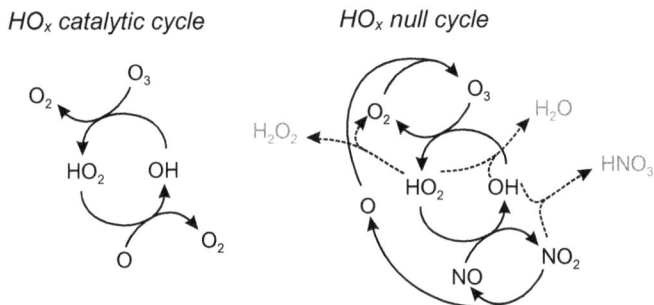

Figure 3.8 HO_x cycles including the catalytic destruction of O_3 (left) and a null cycle (right). The dotted arrows indicate reactions that yield sinks.

Also analogous to the NO_x case, there is a competing null cycle for which there is no net change in the concentration of odd oxygen:

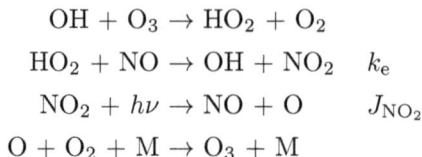

$$OH + O_3 \rightarrow HO_2 + O_2$$
$$HO_2 + NO \rightarrow OH + NO_2 \quad k_e$$
$$NO_2 + h\nu \rightarrow NO + O \quad J_{NO_2}$$
$$O + O_2 + M \rightarrow O_3 + M$$

The steady state between OH and HO_2 is rapidly established, on the timescale of seconds, and hence we can apply the SSA to the OH concentration. In the same way that Equation (3.16) was derived, it is easy to show that the rate of odd oxygen destruction by HO_x reduces to

$$\frac{d[O_x]}{dt} = -2k_d[HO_2][O] \qquad (3.20)$$

i.e. the rate of loss of odd oxygen is twice that of the rate-determining step.

As can be seen in Figure 3.1(b), [O] falls markedly in the lower stratosphere, and the catalytic cycle above becomes increasingly inefficient. Under these conditions, an alternative loss cycle occurs in which ozone directly participates in both steps:

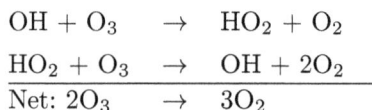

$$
\begin{array}{lcl}
OH + O_3 & \rightarrow & HO_2 + O_2 \\
HO_2 + O_3 & \rightarrow & OH + 2O_2 \\
\hline
\text{Net: } 2O_3 & \rightarrow & 3O_2
\end{array}
$$

Finally, we note that the HO_x family includes not only the hydroxyl, OH, and hydroperoxy, HO_2, radicals, but also H atoms, i.e. $[HO_x] = [H] + [OH] + [HO_2]$. These species are closely coupled, and interconvert on the timescale of seconds. Particularly, noteworthy is the fast termolecular reaction

$$H + O_2 + M \rightarrow HO_2 + M \qquad (3.21)$$

This reaction is two orders of magnitude faster than the reaction $O + O_2 + M$, and ensures that $[HO_2]/[H] \gg 1$, and that [H] is low across most stratospheric altitudes. As a consequence, reactions initiated by H are generally not as important as those involving OH except for those in the upper stratosphere, where H now becomes the dominant form of HO_x (due to the lower pressures and therefore much lower rate of reaction (3.21)). In this region the following catalytic odd oxygen removal process occurs:

$$
\begin{array}{lcl}
H + O_3 & \rightarrow & OH + O_2 \\
OH + O & \rightarrow & H + O_2 \\
\hline
\text{Net: } O_3 + O & \rightarrow & 2O_2
\end{array}
$$

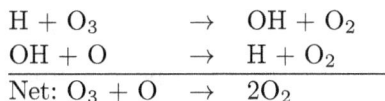

HO_x radicals are removed from the stratosphere by regeneration of H_2O through the reaction

$$OH + HO_2 \rightarrow H_2O + O_2 \tag{3.22}$$

and by reactions to produce the temporary reservoirs HNO_3 and H_2O_2:

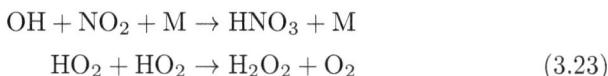

$$OH + NO_2 + M \rightarrow HNO_3 + M$$
$$HO_2 + HO_2 \rightarrow H_2O_2 + O_2 \tag{3.23}$$

Concentrations of HO_x are markedly lower than those of NO_x, at around 10^7 molecules cm^{-3}, and vary with altitude as shown in Figure 3.1(b). Most HO_x is stored in the unreactive HNO_3 form at mixing ratios of ppbv.

OH radicals can be formed in a number of other H-atom abstraction reactions of $O(^1D)$, in addition to the reaction with water described at the start of this section. The most common alternative formation pathway involves proton abstraction from methane, CH_4

$$O(^1D) + CH_4 \rightarrow OH + CH_3 \tag{3.24}$$

The newly formed OH can also react with methane

$$OH + CH_4 \rightarrow H_2O + CH_3 \tag{3.25}$$

Here, CH_4 is acting as an additional source of H_2O, and so of HO_x. As stratospheric levels of both CH_4 and H_2O are rising, stratospheric HO_x levels are set to increase, with major repercussions for ozone loss.

3.2.3 ClO_x Chemistry

The additional O_3 loss rates calculated from these naturally occurring catalytic HO_x and NO_x cycles help reconcile the discrepancy between the $[O_3]$ predicted by the Chapman cycle and measurements of ozone abundance. However, another catalytic cycle of anthropogenic origin, involving Cl atoms and chlorine oxides, provides a further loss mechanism for O_x and leads to stratospheric O_3 depletion, and in particular to formation of the polar O_3 holes which will be discussed in the next section.

CFCs are not found in nature, and anthropogenic emissions of long-lived CFCs such as $CFCl_3$ (known as CFC-11), in the second half of the 20th century have given rise to a third catalytic loss cycle for O_x. CFCs are both photochemically inert in the troposphere and water insoluble, and are efficiently transported to the stratosphere where they can be photolysed by UV radiation at wavelengths below $\sim 255\,nm$ to produce atomic chlorine. These chlorine atoms then acts as the catalyst for an O_3 loss mechanism, analogous to those described above involving NO/NO_2 and OH/HO_2, that involves the cycling between Cl and ClO radicals, known collectively as ClO_x.

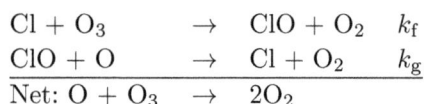

$$
\begin{array}{lll}
Cl + O_3 & \rightarrow \quad ClO + O_2 & k_f \\
ClO + O & \rightarrow \quad Cl + O_2 & k_g \\
\hline
Net: O + O_3 & \rightarrow \quad 2O_2 &
\end{array}
$$

The current CFC mixing ratio is just under 3 ppbv. For comparison, CH_3Cl which is emitted naturally from the oceans has a mixing ratio of 0.6 ppbv. While the CFC flux is low, CFCs have sufficiently long lifetimes (~ 100 years) that their concentration at steady state can be high (see Section 1.6). Since CFCs are stable in the troposphere, this region of the atmosphere acts as a CFC reservoir. As a result, there was a marked increase in stratospheric $[ClO_x]$ throughout the 1980s, in line with Figure 3.9.

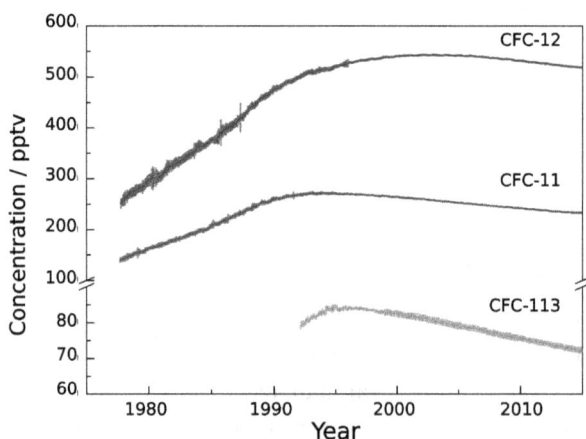

Figure 3.9   Combined CFC-113, -12, and -11 concentration data from the NOAA/ESRL Global Monitoring Division. Note the slow decreases in CFC concentration over the last decade or so.

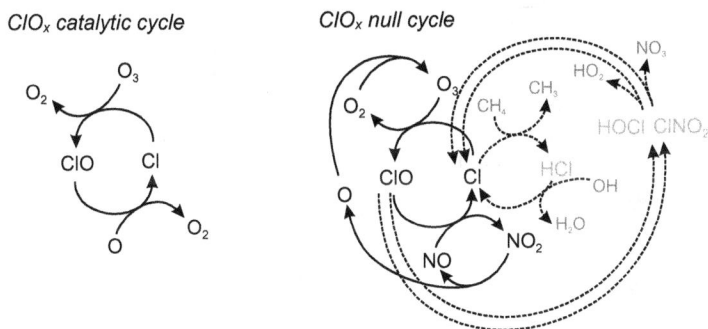

Figure 3.10 ClO$_x$ catalytic destruction of O$_3$ (left) and null cycle (right). The dotted arrows indicate reactions related to temporary reservoirs of Cl.

Analogous to the cases of NO$_x$ and HO$_x$ described previously, the Cl/ClO cycle also competes with the following null cycle, shown in Figure 3.10.

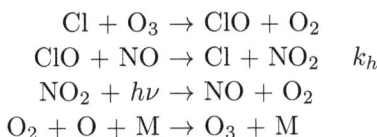

$$Cl + O_3 \rightarrow ClO + O_2$$
$$ClO + NO \rightarrow Cl + NO_2 \quad k_h$$
$$NO_2 + h\nu \rightarrow NO + O_2$$
$$O_2 + O + M \rightarrow O_3 + M$$

As in the NO$_x$ and HO$_x$ cases, the steady state between Cl and ClO is rapidly established, and so we can apply the SSA to the rate of change of [Cl] with the result that the rate of odd oxygen destruction by ClO$_x$ is then

$$\frac{d[O_x]}{dt} = -2k_g[ClO][O] \tag{3.26}$$

Catalytic ClO$_x$ cycles are terminated by conversion of ClO$_x$ into the non-radical chlorine reservoir species HCl, ClONO$_2$, and HOCl:

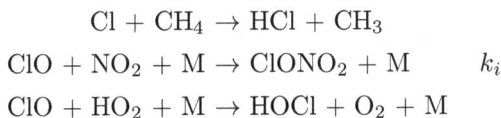

$$Cl + CH_4 \rightarrow HCl + CH_3$$
$$ClO + NO_2 + M \rightarrow ClONO_2 + M \quad k_i$$
$$ClO + HO_2 + M \rightarrow HOCl + O_2 + M$$

These reservoirs, however, are temporary, and eventually ClO$_x$ is regenerated, thereby reinitiating the catalytic ClO$_x$ cycle:

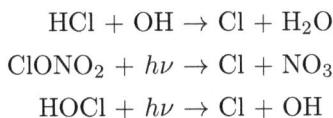

$$HCl + OH \rightarrow Cl + H_2O$$
$$ClONO_2 + h\nu \rightarrow Cl + NO_3$$
$$HOCl + h\nu \rightarrow Cl + OH$$

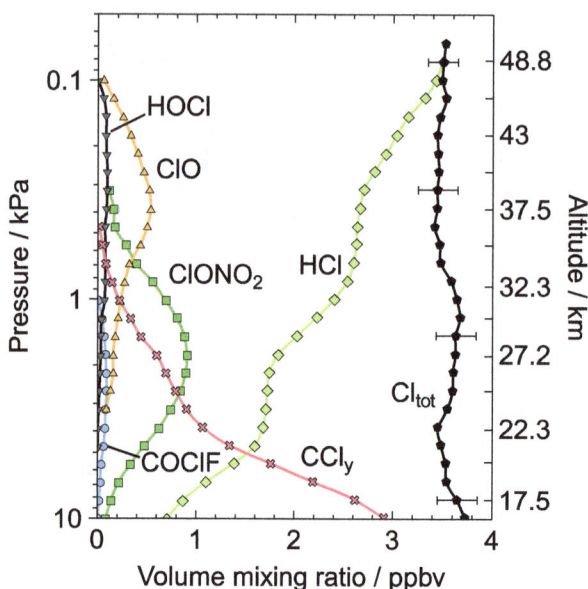

Figure 3.11   Volume mixing ratios of ClO_x reservoir species as a function of altitude. *Source*: Adapted from R. Zander *et al.*, *Geophys. Res. Lett.* **23** 2357 (1996).

The volume mixing ratio of the chlorine reservoir compounds and their altitude dependence is displayed in Figure 3.11. In common with many of the species discussed so far, ClO_x, HCl, $ClONO_2$, and HOCl (collectively known as Cl_y) are ultimately removed from the stratosphere by transport to the troposphere and subsequent deposition. We note that the reaction $ClO + NO_2 \rightarrow ClONO_2$ couples the Cl_x and NO_x cycles, and at high stratospheric [Cl] can lead to nonlinear responses of ozone concentrations to Cl perturbations. It should also be noted the different role that OH plays here — in NO_x cycles, OH acts to form the reservoir species HNO_3, whereas in ClO_x cycles, OH acts to release Cl from the HCl reservoir.

To address the phenomenon of large-scale stratospheric O_3 loss, particularly in the polar regions, caused by the increased levels of chlorine-containing compounds generated by anthropogenic emissions, international agreements have been adopted and implemented to reduce the consumption and production of CFCs, halons, and trichloroethane, and also hydrochlorofluorocarbons (HCFCs), the first generation of CFC replacement gases.[3]

---

[3]HCFCs were thought to be attractive alternatives to CFCs, as they contain a C–H bond and can react with OH in the troposphere, making the Cl they contain less likely

The *Montreal Protocol* on ozone-depleting substances was adopted in September 1987 and banned the use of CFCs and replacement substances also recognised as ozone-depleting pollutants, and has so far been ratified by some 190 parties. As a result, the rate of stratospheric ozone depletion has slowed considerably since its implementation. While chlorine concentrations in the stratosphere are declining, we note that the decline is slow, partly because CFC concentrations are decreasing more slowly than expected, and partly because the quantities of HCFCs that were used to replace them are still rising despite bans on production and use, due to the long atmospheric lifetimes of these molecules. Because of the long lifetimes of these species, they will continue to play an important, but diminishing, role in ozone chemistry during the present century.

3.2.4 Bromine Chemistry

Given that Cl radicals are extremely effective at removing ozone, it is pertinent to consider whether any other halogens might also be efficient O_3 destroyers. The relative efficiencies of F, Cl, and Br in stratospheric ozone destruction on a *molecule-for-molecule* basis is in the order Br > Cl > F. This order reflects the proportion of radicals present in *active* form rather than in reservoirs, as the reactivities of the different halogen radicals are similar. Firstly, a radical cycle with fluorine is not important because of the high stability of the HF bond; effectively, fluorine is always found in its unreactive reservoir form. In the case of iodine, there are insufficient I-containing species for iodine to be a significant ozone destroyer. Bromine however is a significant O_3 destroyer and is present at low concentrations of $\sim$20 pptv. Its major natural source is CH_3Br from vegetation, while anthropogenic sources arise from soil fumigation and fire retardants. The Br catalytic cycle is the same as for Cl:

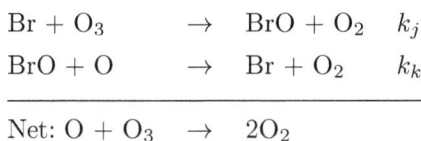

$$\begin{array}{lll} Br + O_3 & \rightarrow & BrO + O_2 \quad k_j \\ BrO + O & \rightarrow & Br + O_2 \quad k_k \\ \hline Net: O + O_3 & \rightarrow & 2O_2 \end{array}$$

The key difference between Br and Cl in the atmosphere is that Br has no efficient reservoirs. Thus, despite the low concentration of Br, most of it is

---

to reach the stratosphere. However, HCFCs and HFCs (halogen free replacements for CFCs) are also potent greenhouse gases (GHGs), and their production is now regulated.

active and so, on a per atom basis, it is a more powerful ozone destroyer than Cl. For example, the reaction of Br with CH_4 to form HBr is endothermic and therefore very slow and not important in the atmosphere. Instead, the major source of HBr, the dominant Br reservoir, is the reaction with HO_2.

$$Br + HO_2 \rightarrow HBr + O_2 \tag{3.27}$$

This reaction rate is slow due to low $[HO_2]$. Another possible reservoir species is $BrONO_2$, formed and destroyed by the reactions $BrO + NO_2 + M \rightarrow BrONO_2 + M$ and $BrONO_2 + h\nu \rightarrow Br + NO_3$, respectively. However, as $BrONO_2$ photolysis occurs at longer wavelengths than for $ClONO_2$, and has a larger absorption cross-section than $ClONO_2$, then $BrONO_2$ photolysis happens much more rapidly than $ClONO_2$ photolysis, and makes this reservoir inefficient. As shown in Figure 3.12, the major BrO_x species is active BrO; this should be compared with ClO_x shown in Figure 3.11, for which the dominant species is reservoir HCl. In summary, BrO_x tends to be present as active BrO and Br species, rather than in reservoir species thereby making Br a potent ozone destroyer. It should be emphasised however that as $[Cl] \gg [Br]$, the significance of the halogens to O_3 destruction varies in the order Cl > Br > F.

3.2.5 Summary of the Catalytic Cycles

We conclude this section by summarising the important points from the previous sub-sections (see Figure 3.13). Most importantly, the generic

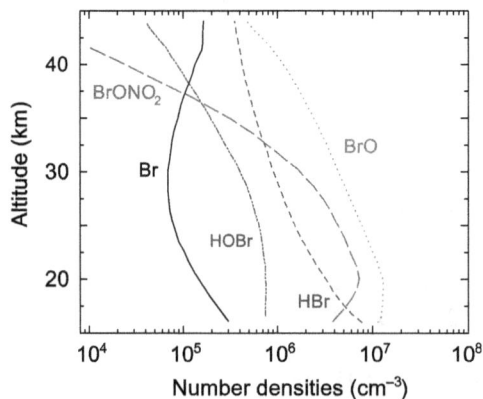

Figure 3.12   Concentration of Br containing species in the stratosphere as a function of altitude.

Source: Adapted from Y.L. Yung *et al.*, *J. Atmos. Sci.* **37** 339 (1980).

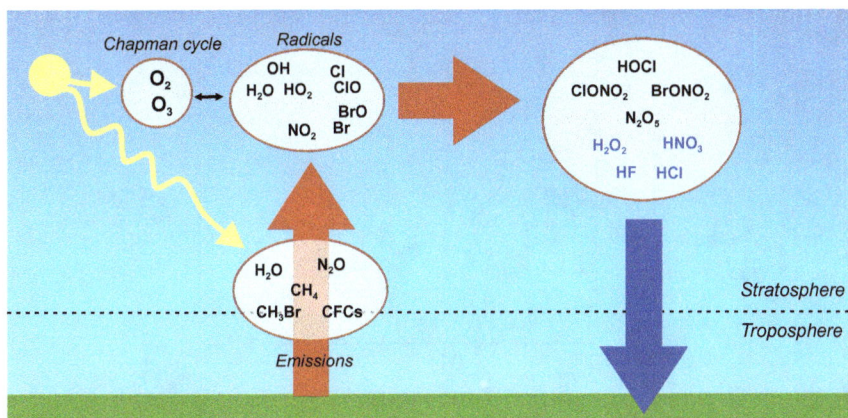

Figure 3.13   Summary of catalytic species, their sources, and their reservoirs.

catalytic odd oxygen removing reaction cycle is:

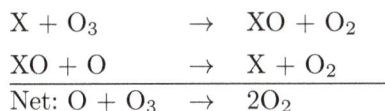

$$
\begin{array}{lcl}
X + O_3 & \rightarrow & XO + O_2 \\
XO + O & \rightarrow & X + O_2 \\
\hline
\text{Net: } O + O_3 & \rightarrow & 2O_2
\end{array}
$$

With a null cycle given by:

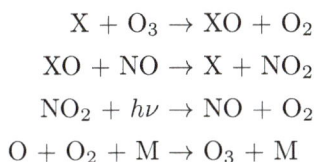

$$X + O_3 \rightarrow XO + O_2$$
$$XO + NO \rightarrow X + NO_2$$
$$NO_2 + h\nu \rightarrow NO + O_2$$
$$O + O_2 + M \rightarrow O_3 + M$$

The cycles we have to consider involve O_x, NO_x, HO_x, ClO_x, and BrO_x, and in almost all cases, the rate-determining step is the reaction with O, i.e. $O + NO_2$, $O + ClO$, $O + HO_2$. Furthermore, catalytic cycles involving members of different families occur. For example, ClO_x and BrO_x are coupled by the reaction $ClO + BrO \rightarrow Cl + Br + O_2$, and changes in the abundances of one family will result in changes in the others. Thus, the system is nonlinear, although reasonably well behaved.

The relative importance of a given cycle varies with altitude as shown in Figure 3.14. The importance of each cycle is dictated by a combination of: (i) the magnitude of the rate constants involved; (ii) the lifetimes of relevant

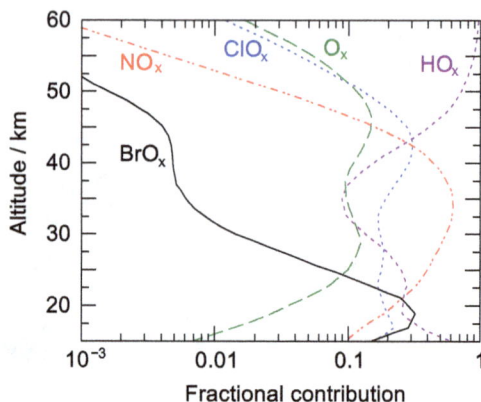

Figure 3.14   The fractional contribution of each catalytic cycle to ozone loss as a function of altitude. Adapted from *Atmospheric Chemistry* by A.M. Holloway and R.P. Wayne, Royal Society of Chemistry, Cambridge (2010). Original figure kindly provided by R.P. Wayne (Oxford).

source gases; and (iii) the partitioning between reservoir and active species. The rates of all catalytic cycles are seen to decrease as altitude decreases. To appreciate why this is so, consider the NO_x cycle. The rate depends on both $[NO_2]$ and $[O]$, both of which vary with altitude. At low altitudes the dominant nitrogen species are reservoir species, with the active NO and NO_2 species only dominating at altitudes above 30 km. More importantly, $[O]$ also increases with increasing altitude; this is because the rate of formation of O via photolysis of O_2 increases with altitude as the flux of short-wavelength radiation increases. In addition, the rate of the reaction that destroys oxygen atoms, $O + O_2 + M \rightarrow O_3 + M$, decreases with increasing altitude as the total atmospheric pressure falls. Finally, the total rate of loss of odd oxygen loss is given by:

$$d[O_x]/dt = 2J_1[O_2] - 2k_4[O_3][O] - 2k_b[NO_2][O] - 2k_d[HO_2][O]$$
$$-2k_g[ClO][O] - 2k_k[BrO][O]$$

3.3 The Ozone Hole: The Importance of Heterogeneous Chemistry

3.3.1 O_3 Column Amounts over the Antarctic

From the discussion in Section 3.1, it is clear that while the rate of ozone formation is a maximum at the equator, the ozone distribution is

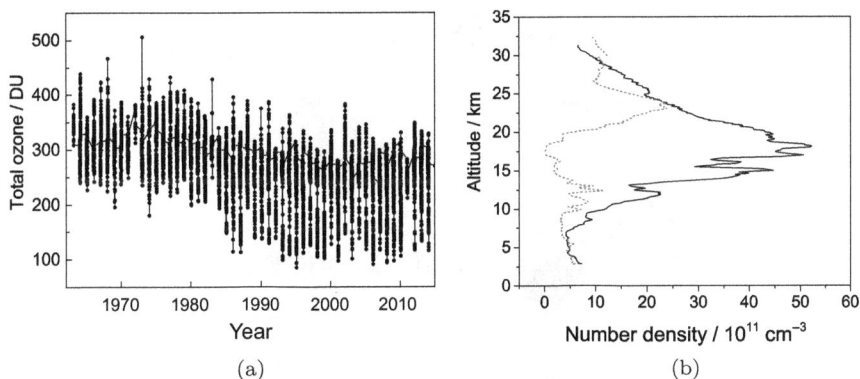

Figure 3.15 (a) Measurements of total ozone over the South Pole obtained using a Dobson spectrophotometer (http://www.esrl.noaa.gov/gmd/dv/site/), (b) vertical ozonesonde profile in August (solid lines) and October (dotted lines) 1993 (ftp://aftp.cmdl.noaa.gov/data/ozwv/).

affected by transport. In the following discussion, we will consider the ozone column in the Antarctic. Figure 3.15(a) shows the mean annual column amount of ozone over a 50 year period, as reported by the British Antarctic Survey using a Dobson ozone spectrophotometer located at Halley station (at a latitude of $-76°$). Up until the mid-to-late 1970s, the average total ozone column was around 300 DU, with a minimum of about 250–300 DU in spring and a maximum of about 400 DU in summer. However, during the 1980s a dramatic decrease in the column was observed. Furthermore, this loss is seasonal, starting in spring and recovering in early summer, i.e. the loss develops rapidly over a period of around 6 weeks. Indeed, as shown in Figure 3.15(b), there is near total loss of ozone over the 15–20 km altitude range during the Antarctic spring. These losses are recurrent and have been confirmed by satellite observations, as depicted in Figure 3.16. Later in the year, ozone returns to the Antarctic stratosphere, although the region of depleted ozone may be transported to lower latitudes before this recovery is complete. This large-scale loss of ozone is widely known as the *ozone hole* — typically defined when the ozone column amount < 220 DU — and occurs with a spatial scale that is similar to the size of the Antarctic continent.[4] These observations of

---

[4]Interestingly, ozone loss is also observed in the Arctic spring but not to such a large extent. Arctic ozone loss also displays larger inter-annual variability than the Antarctic loss.

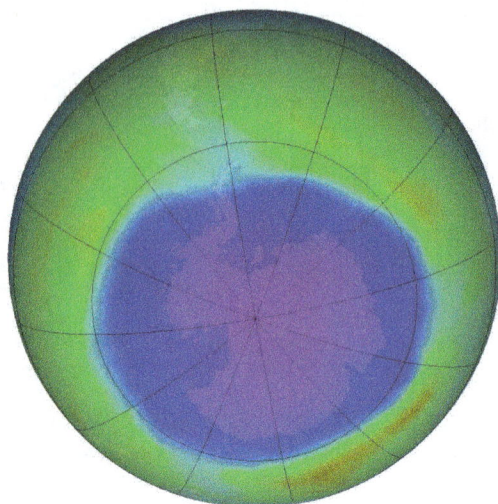

Figure 3.16   Satellite measurements of polar O_3 levels; blue denotes a deficit of O_3.

Source: Data from NASA (http://earthobservatory.nasa.gov/IOTD/view.php?id=86869 &eocn=image&eoci=related_image), 2 October 2015.

large ozone losses in the lower stratosphere during the Antarctic spring are inexplicable in terms of the gas-phase ClO_x chemistry that was considered in Section 3.2.3. Instead, a new mechanism based upon a mixture of transport and heterogeneous reactions on the surfaces of cloud particles is necessary.

3.3.2 Explaining Polar Ozone Losses

CFCs are prime candidates to be the cause of polar O_3 loss. As shown in Figure 3.9, their abundance in the atmosphere grew markedly during the 1970s, with concomitant increases in Antarctic ClO_x levels. However, because the rate-determining step in the ClO_x mechanism depends on the concentration of atomic oxygen, which is low in the stratosphere, the formation of the ozone hole cannot be explained by the relatively slow gas-phase chemistry described in Section 3.2.3. A key piece of evidence for elucidating a suitable alternative mechanism to explain the ozone hole came from the observed *anticorrelation* of $[O_3]$ and $[ClO]$ as a function of latitude, shown in Figure 3.17. Furthermore, $[ClO]$ is unexpectedly high, as it is expected to reside predominantly as the HCl and $ClONO_2$ reservoir

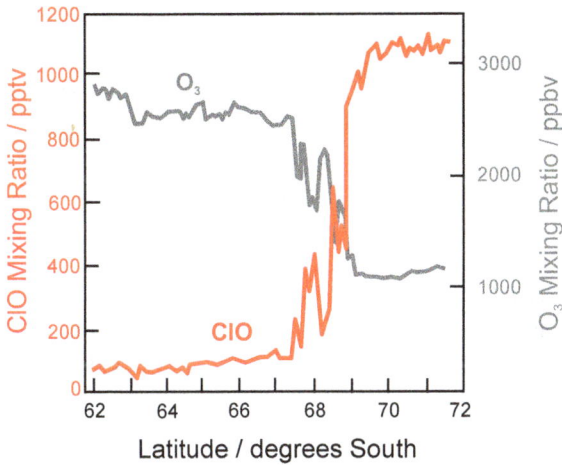

Figure 3.17 The anticorrelation of O_3 and ClO mixing ratios as a function of latitude. *Source*: Adapted from J.G. Anderson *et al.*, *J. Geophys. Res.* **94** 11465 (1989).

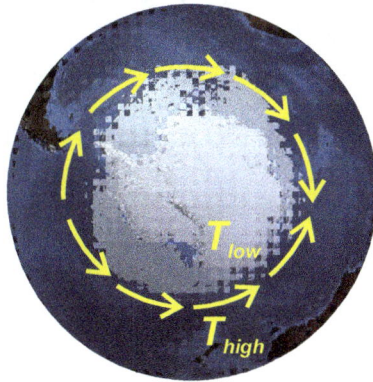

Figure 3.18 Creation of the polar vortex over the South Pole.

species. We also note that $[NO_x]$ was also found to be depleted at high latitudes.

In winter, the polar regions receive no sunlight, but still emit infrared (IR) radiation to space. At the same time, a strong circumpolar vortex forms in the Antarctic stratosphere. This prevents latitudinal mixing of air, as shown in Figure 3.18 (also see Section 1.5), and as a consequence there is little latitudinal transport of chemicals into or out of the vortex, which therefore acts like an isolated chemical reactor. The result of having well isolated air within the vortex is that the temperatures inside can drop as low as $180\,K$. At these low temperatures, clouds form in the lower stratosphere up to altitudes of $30\,km$. These *polar stratospheric clouds* (PSCs) contain both H_2O and HNO_3, and form on liquid sulphate aerosols. H_2O and HNO_3 co-condense to form nitric acid trihydrate (NAT), $HNO_3 \cdot 3H_2O$. PSCs can form at slightly higher temperatures than pure H_2O clouds, meaning that they form in greater volume than 'normal' clouds. PSCs provide surfaces for catalysis, allowing reactions to occur that would not occur in the gas phase. Most importantly these surfaces allow the efficient conversion of ClO_x reservoir species into active Cl via the reaction $HCl_{(s)} + ClONO_{(s)} \rightarrow Cl_{2(g)} + HNO_{(s)}$. The Cl_2 product is then photolysed on the return of sunlight to yield Cl radicals which can then react with O_3 to produce ClO via the reaction $Cl + O_3 \rightarrow ClO + O_2$.[5] The reaction on the surface of the PSC is far faster than the gas-phase reactions that convert HCl and $ClONO_2$ into Cl_2. In fact, the reaction is so fast that it can essentially be regarded as quantitative, such that one or other reagent is completely titrated. Under these physical conditions, HCl and $ClONO_2$ no longer act as effective reservoirs for ClO_x, meaning that a greater proportion of chlorine remains as ClO_x, including ClO (albeit in the form of Cl_2 during the winter).

The main loss process for ClO is via the reaction $ClO + NO_2 + M \rightarrow ClONO_2 + M$. However, as noted above, $[NO_2]$ is low when $[ClO]$ is high, and so this loss process is slow; reactions to destroy active Cl are therefore also very slow. The low value of $[NO_2]$ is a consequence of the fact that the HNO_3 formed in reactions on PSCs can be permanently removed by sedimentation in a process known as *denitrification*. Hence, as the destruction of ClO is slow, proportionally more ClO is present. This high $[ClO]$ is maintained, as the isolated vortex ensures that $[NO_x]$ remains low (there is no new input of NO_x from transport). Thus, both water vapour concentration and $[NO_x]$ are low when $[ClO]$ is high. As both H_2O and HNO_3 condense to form NAT particles, both are removed from the stratosphere during gravitational settling. PSCs therefore do two things: (i) convert Cl reservoir compounds into ClO_x; and (ii) remove HNO_3 by sedimentation, so that there is slow recovery of ClO_x back into reservoirs.

---

[5]It should be noted that these steps are stoichiometric, not catalytic, and result in a small ozone loss of *ca.* 0.05%.

Once built up to a sufficient level, ClO can then destroy ozone via the following catalytic cycle:

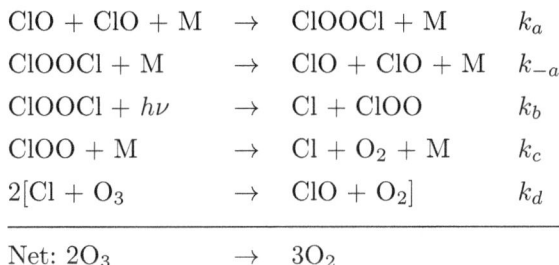

$$
\begin{array}{llll}
\text{ClO} + \text{ClO} + \text{M} & \rightarrow & \text{ClOOCl} + \text{M} & k_a \\
\text{ClOOCl} + \text{M} & \rightarrow & \text{ClO} + \text{ClO} + \text{M} & k_{-a} \\
\text{ClOOCl} + h\nu & \rightarrow & \text{Cl} + \text{ClOO} & k_b \\
\text{ClOO} + \text{M} & \rightarrow & \text{Cl} + \text{O}_2 + \text{M} & k_c \\
2[\text{Cl} + \text{O}_3 & \rightarrow & \text{ClO} + \text{O}_2] & k_d \\
\hline
\text{Net: } 2\text{O}_3 & \rightarrow & 3\text{O}_2 &
\end{array}
$$

Within this kinetic scheme the rate-determining step for O_3 loss is the photolysis of ClOOCl; this is because of the competition between photolysis and thermal decomposition of the dimer. At higher temperature, decomposition is more rapid and competes with photolysis, and the cycle is less efficient. Ignoring thermal decomposition, it can be shown that $d[O_3]/dt = -2k_a[\text{ClO}]^2[\text{M}]$, and combining this result with measured rate constants and typical concentration values yields an estimated time constant for ozone loss of ~ 60 days, consistent with observations. While the catalytic cycle above predicts the correct order of magnitude for the ozone loss, it does underestimate the loss somewhat. Better agreement with observation is obtained when cycles involving Br are also considered; these are found to account for 30% of the total loss. Most importantly, we note the following two points about this new mechanism. Firstly, O atoms are not involved at all, and so the loss mechanism can operate at *any* altitude. Secondly, as ozone itself is not involved in the rate-determining step, *total* removal of ozone can occur. Finally, mixing ratios of stratospheric chlorine increased by about a factor of 3 between 1970 and 1990. As $d[O_3]/dt \propto [\text{ClO}]^2$, this led to an order of magnitude increase in O_3 loss, accounting for the sudden appearance of the Antarctic ozone hole.

HNO_3 is a by-product of PSC reactions, and remains as particles at low temperature. In the Antarctic summer, when the stratosphere warms, PSCs evaporate and so $[\text{HNO}_3]$ increases. This gaseous HNO_3 acts as a source of NO_2, either by direct photolysis, $\text{HNO}_3 + h\nu \rightarrow \text{OH} + \text{NO}_2$, or by reaction, $\text{HNO}_3 + \text{OH} \rightarrow \text{H}_2\text{O} + \text{NO}_3$ followed by, $\text{NO}_3 + h\nu \rightarrow \text{O} + \text{NO}_2$. The resulting NO_2 then reacts with ClO to form the reservoir species $ClONO_2$. As $[\text{ClO}]$ is low there is then less ozone destruction, and in time ozone can begin to be reformed. The timescale for this destruction and subsequent

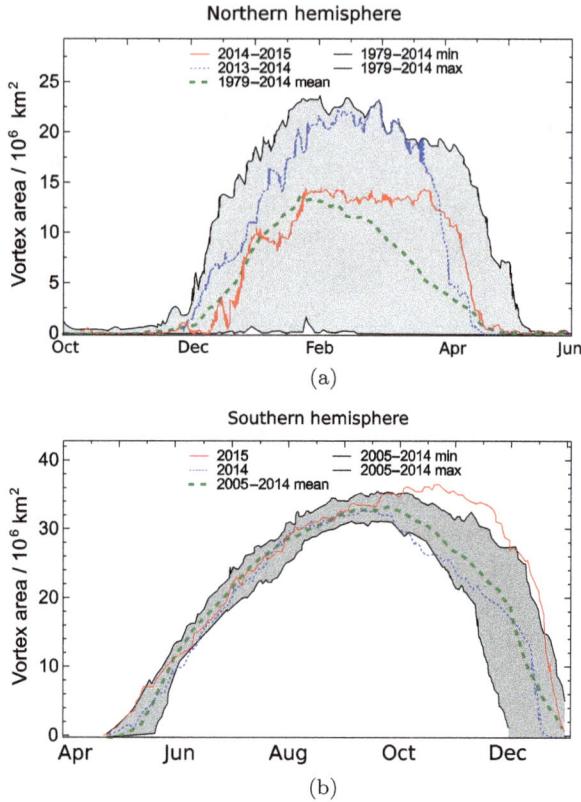

Figure 3.19   Spinning up of the polar vortex in the (a) Northern and (b) Southern hemisphere.

Source: Data taken from http://www.cpc.ncep.noaa.gov/products/stratosphere/polar/ polar.shtml.

recovery cycle is shown in Figure 3.19, which plots the spatial extent of the polar vortex over a 9-month period.

Ozone loss in the Arctic occurs by the same mechanism, but the ozone hole there is less extended. This is because the winter polar vortex in the Northern hemisphere is more disturbed dynamically (because of the Greenland land mass) and there is greater heat transport from lower latitudes. As shown in Figure 3.20, temperatures are warmer than in the Southern hemisphere by approximately 10 K on average, and are also more variable. The very cold winter conditions that allow PSCs to form are therefore less common in the Arctic stratosphere. Furthermore, the higher temperatures mean that the thermal decomposition of ClOOCl is faster

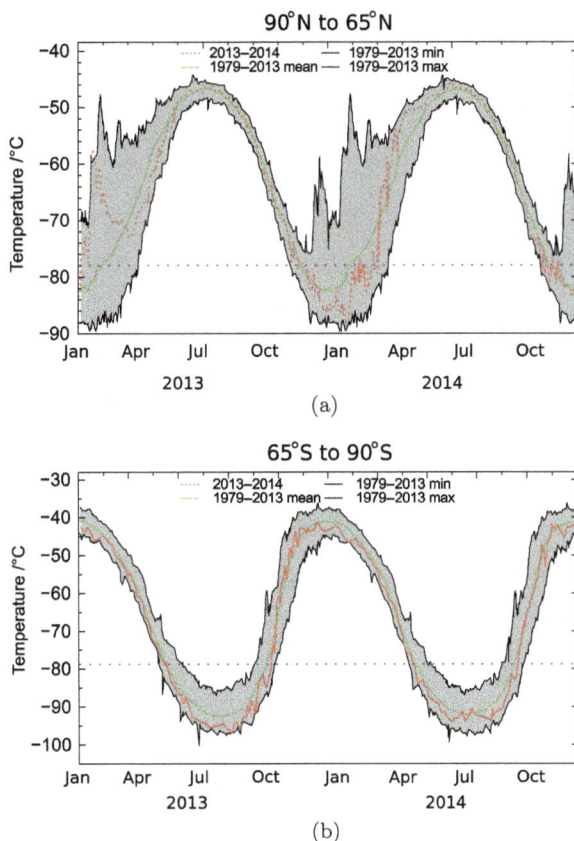

Figure 3.20   The temperature of the stratosphere in (a) the Arctic and (b) Antarctic with the PSC formation temperature indicated by the dotted line.

Source: Data taken from http://www.cpc.ncep.noaa.gov/products/stratosphere/polar/polar.shtml.

in the Northern hemisphere, so reducing the effectiveness of the ClOOCl catalytic cycle. In addition, the increased mixing with lower-latitude air that contains high levels of $[NO_2]$ lowers $[ClO]$ in the Arctic via the reaction $ClO + NO_2 + M \rightarrow ClONO_2 + M$. While $[ClO]$ levels in the Arctic are still relatively high, with levels consistent with release of Cl_2 from the PSC reaction $ClONO_{2(g)} + HCl_{(s)} \rightarrow Cl_{2(g)} + HNO_{3(s)}$, the rate of removal of HNO_3 by sedimentation is lower, and so there is more NO_2 available and therefore a shorter recovery time as $[ClO]$ falls back to low values. In short, in the Northern hemisphere there is less effective destruction occurring for a shorter period of time, resulting in lower total ozone loss.

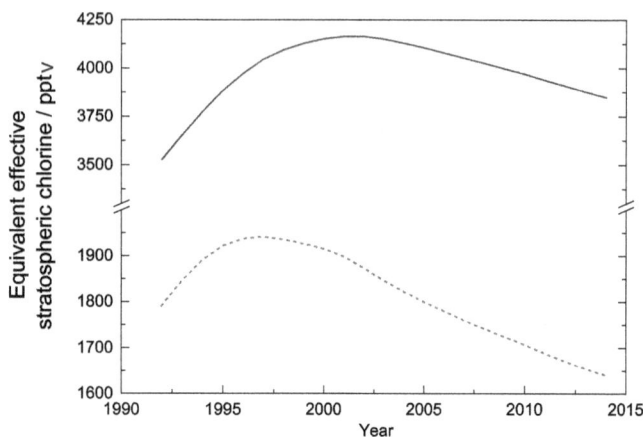

Figure 3.21   EESC at mid-latitudes (dashed line) and over Antarctica (solid line).
Source: Data taken from http://www.esrl.noaa.gov/gmd/odgi/.

3.4 Future of Stratospheric Ozone Due to Cooling by Climate Change

As mentioned above, CFC emissions have been dramatically reduced due to the Montreal protocol. The *effective equivalent chlorine in the stratosphere*, EESC, has reached a maximum and is now falling, as shown in Figure 3.21. EESC includes both Cl and Br chemistry, and natural and anthropogenic chlorine; natural chlorine comes from CH_3Cl at the 0.5 ppbv level, while the anthropogenic EESC input is $\sim$3 ppbv. If one assumes that nothing else has changed then the ESSC level will return to pre-1980s levels by 2050, i.e. with no other atmospheric changes, the Antarctic hole will have recovered by 2050. Even so, by 2050 the EESC levels will not have fallen below the 1980 EESC values, which are taken as the threshold for the appearance of the O_3 hole. The timescale for the eventual decay of EESC levels below the threshold is determined by the hundred-year-or-more lifetimes of CFC-11 and CFC-12.

The Montreal protocol represents a real success story in terms of our influence on the atmosphere and understanding of atmospheric chemistry, and the ozone hole is consequently well on the way to repairing itself. However, we certainly cannot afford to be complacent. Climate change is very much a reality, and increased $[CO_2]$ means a cooler stratosphere (see Section 2.3), with a variety of potential consequences for ozone destruction. Firstly, stratospheric cooling leads to increased formation of PSCs and so enhanced polar ozone losses. This effect delays recovery of the ozone hole and explains why record losses have recently been observed despite elimination of CFC emissions. Stratospheric cooling therefore has the consequence that the future upper limit on EESC required to eliminate the formation of an ozone hole will actually be lower than before. This effect may be offset to some extent by increases in polar transport of O_3 from mid-latitudes.

Increased formation of PSCs is not the only consequence of stratospheric cooling. In the mid- to upper stratosphere, cooling will slow the rates of gas-phase reactions that destroy O_3, while in the lower stratosphere, the effect will be to convert HO_x into reservoir compounds. Interconversion between active OH and NO_2 and the HNO_3 reservoir is achieved in the lower stratosphere by the reactions $OH + NO_2 + M \rightarrow HNO_3 + M$ and $HNO_3 + OH \rightarrow H_2O + NO_3 + M$. The two reactions convert HO_x into a reservoir species and water, respectively, and both are faster at lower temperatures. This leads to an overall reduction in HO_x, and is predicted to lead to an increase in lower stratospheric O_3 levels later this century, an effect known as *super-recovery*. It should however be noted that stratospheric/tropospheric exchange is a source of tropospheric O_3, and so tropospheric O_3 levels are also likely to increase with climate change — the role of ozone in tropospheric chemistry is one of the focal points of the next chapter.

3.5 Questions

3.5.1 Essay-style Questions

Q3.1: Which wavelengths of solar radiation are absorbed by O_3 and O_2? Write out the four main reactions that constitute the Chapman mechanism for the ozone layer. In practice, the Chapman reaction scheme predicts significantly greater stratospheric ozone concentrations than those determined by direct measurement. Suggest an explanation for this discrepancy.

Q3.2: What is the main source for stratospheric NO_x? Detail the characteristics of molecules that are emitted at the Earth's surface and survive to reach the stratosphere where they are sources of NO_x, ClO_x, etc.

Q3.3: What are the reactions that constitute the catalytic cycle for ozone depletion by chlorine? How does NO_x impact the efficiency of this catalytic cycle? What are the primary reservoir species for chlorine in the stratosphere?

Q3.4: What is the *ozone hole* and where and when does it occur? In your answer, detail the special meteorological conditions that allow the ozone hole to form. Also state the specific changes in chlorine chemistry that occur to form the ozone hole and highlight the role that nitrogen compounds play in ozone hole formation.

Q3.5: Contrast the Arctic and Antarctic polar stratospheres and how this affects the occurrence of the ozone hole.

Q3.6: Explain why Br is more damaging to ozone than Cl, and why F does not destroy ozone.

3.5.2 Problems

P3.1: In the atmosphere, molecular oxygen is photochemically decomposed as follows:

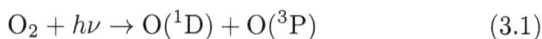

$$O_2 + h\nu \rightarrow O(^1D) + O(^3P) \tag{3.1}$$

The major loss processes for these atoms are:

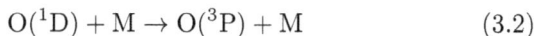

$$O(^1D) + M \rightarrow O(^3P) + M \tag{3.2}$$

and

$$O(^3P) + O_2 + M \rightarrow O_3 + M \tag{3.3}$$

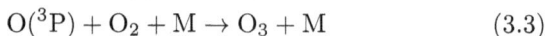

with rate constants k_q and k_t respectively and where M is N_2 or O_2.

(a) By equating the rate of reaction (3.1) to I_{abs} and noting reaction (3.2), write down a kinetic equation for $d[O(^1D)]/dt$.

(b) Using the result in (a) find an approximate expression for $[O(^1D)]$ by applying the steady-state hypothesis to $O(^1D)$.

(c) By integration of the rate law in (a) show that

$$[O(^1D)] = \left(\frac{I_{abs}}{k_q[M]}\right)(1 - \exp(-k_q[M]t))$$

assuming constant irradiation is applied after $t = 0$ and that no $O(^1D)$ exists prior to this time. Under what conditions does this expression correspond to the approximate result derived in (b)?

(d) In the atmosphere at an altitude of *ca.* 80 km, $[M] \approx 3 \times 10^{14}$ molecules cm^{-3} and the composite k_q for N_2 and O_2 is 3×10^{-11} cm^3 molecule^{-1} s^{-1}. Estimate the minimum illumination time required for the non-steady-state and steady-state concentrations to be identical to within 1%. Is the steady-state hypothesis a good approximation for the atmospheric behaviour of $O(^1D)$ where the solar intensity changes over periods of hours?

(e) Neglecting reaction (3.2), give steady-state and non-steady-state expressions for $[O(^3P)]$ in terms of I_{abs}, k_t, $[M]$, and $[O_2]$ assuming the latter concentrations are in excess.

(f) Under atmospheric conditions corresponding to part (d), $k_t \approx 1.4 \times 10^{-33}$ cm^6 molecule^{-2} s^{-1}. Is the steady-state hypothesis a suitable approximation to apply to $O(^3P)$ in atmospheric modelling?

P3.2: The important "oxygen-only" reactions in the stratosphere can be written as:

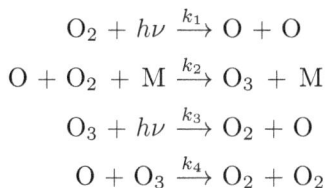

$$O_2 + h\nu \xrightarrow{k_1} O + O$$
$$O + O_2 + M \xrightarrow{k_2} O_3 + M$$
$$O_3 + h\nu \xrightarrow{k_3} O_2 + O$$
$$O + O_3 \xrightarrow{k_4} O_2 + O_2$$

(a) Under conditions of constant illumination, it is possible to write the rates of O atom production in photolysis steps (1) and (3) in the form $2k_1[O_2]$ and $k_3[O_3]$ where k_1 and k_3 have units of s^{-1}. Describe the factors which are likely to determine the magnitudes of k_1 and k_3.

(b) Using the photolysis rate expressions in (a), write down the rate equations for $[O]$ and $[O_3]$.

(c) Under the conditions described above, use the SSA to show that

$$[O][O_3] = \frac{k_1[O_2]}{k_4}$$

With the help of this result and the rate equation for $[O_3]$ obtained in (b), derive the following steady-state expression for the ozone concentration

$$[O_3] = [O_2]\frac{k_1}{2k_3}(x - 1)$$

where

$$x = \sqrt{1 + \frac{4k_2k_3[M]}{k_1k_4}}$$

P3.3: In this problem, we consider the Chapman mechanism at altitudes of 20 and 45 km where the atmospheric temperature is 200 and 270 K, respectively. The total number densities are 1.8×10^{18} and 4.1×10^{16} molecules cm^{-3}, and the volume mixing ratios of O_3 are 2.7 and 3.3 ppmv, at 20 and 45 km respectively. We also assume that

(i) O_3 photolysis occurs with a rate constant $J = 10^{-2}$ s^{-1}, that
(ii) O_3 formation occurs with a rate constant of 10^{-33} cm^6 molecule^{-2} s^{-1}, and that
(iii) the reaction of O with O_3 occurs with a rate constant $A\exp(-2030/T)$ where $A = 8 \times 10^{-12}$ cm^3 molecule^{-1} s^{-1}.

(a) Use the Chapman mechanism to calculate the lifetime of O atoms, τ_O, at both altitudes. Can O atoms be assumed to be in steady state throughout the stratosphere?
(b) Use Equation (3.3) to calculate the ratio $[O]/[O_3]$ at 20 km and 45 km. Can it be assumed that $[O_3] = [O_x]$ throughout the stratosphere?
(c) Use Equation (3.4) to calculate the lifetime of O_x, τ_{O_x}, at 20 and 45 km. In which part of the stratosphere do you expect O_x to be in chemical steady state?

P3.4: In considering the effects of photolysis, explain why the rate of ozone production at altitude z is taken to be proportional to $n(z)I(z)$, where $n(z)$ is the number density of O_2 molecules and $I(z)$ is the intensity of the photolysing radiation. Given that the absorption cross-section is $\sigma = 4 \times 10^{-23}$ cm^2 molecule^{-1} and that $n(z) = n_0 \exp(-z/H)$ where $n_0 = 5 \times 10^{18}$ molecules cm^{-3} and $H = 7$ km, obtain an expression for dI/dz and hence estimate the altitude at which the rate of ozone production is a maximum.

P3.5: At altitudes near and above 30 km, the following reactions significantly affect the chemistry of O_3:

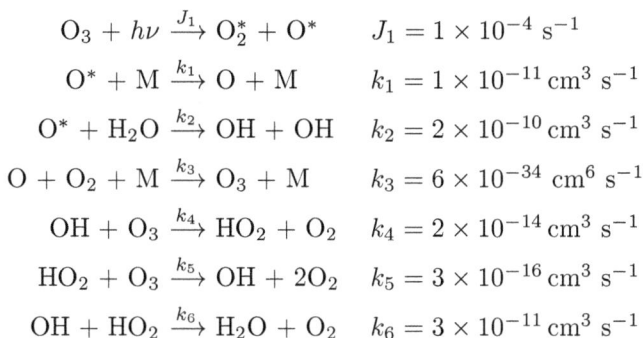

$$O_3 + h\nu \xrightarrow{J_1} O_2^* + O^* \qquad J_1 = 1 \times 10^{-4} \text{ s}^{-1}$$

$$O^* + M \xrightarrow{k_1} O + M \qquad k_1 = 1 \times 10^{-11} \text{ cm}^3 \text{ s}^{-1}$$

$$O^* + H_2O \xrightarrow{k_2} OH + OH \qquad k_2 = 2 \times 10^{-10} \text{ cm}^3 \text{ s}^{-1}$$

$$O + O_2 + M \xrightarrow{k_3} O_3 + M \qquad k_3 = 6 \times 10^{-34} \text{ cm}^6 \text{ s}^{-1}$$

$$OH + O_3 \xrightarrow{k_4} HO_2 + O_2 \qquad k_4 = 2 \times 10^{-14} \text{ cm}^3 \text{ s}^{-1}$$

$$HO_2 + O_3 \xrightarrow{k_5} OH + 2O_2 \qquad k_5 = 3 \times 10^{-16} \text{ cm}^3 \text{ s}^{-1}$$

$$OH + HO_2 \xrightarrow{k_6} H_2O + O_2 \qquad k_6 = 3 \times 10^{-11} \text{ cm}^3 \text{ s}^{-1}$$

where O^* and O_2^* are electronically-excited metastable states of atomic and molecular oxygen respectively. OH and HO_2 are collectively known as odd hydrogen, HO_x. At 30 km, the density of the atmosphere is about 5×10^{17} cm^{-3} and the mole fractions of water vapour and ozone are each 2×10^{-6}.

(a) Discuss methods by which OH and HO_2 may be monitored in the atmosphere.
(b) What are the likely term symbols for O^* and O_2^*?
(c) Why are O^* and O_2^* metastable?
(d) Calculate the approximate steady-state concentration of O^*.
(e) Given that HO_x is in steady state, show that

$$[OH][HO_2] = \frac{k_2}{k_6}[O^*][H_2O]$$

Hence, by also assuming that the reactions associated with k_4 and k_5 occur at the same rate, calculate the steady-state concentrations of OH and HO_2.

Chapter 4

Tropospheric Chemistry

Introduction

The previous chapter highlighted the chemistry involved in formation of the ozone layer, and the importance of this layer, together with stratospheric O_2, in shielding the Earth's surface from harmful ultraviolet radiation. The highly effective filtering of the actinic flux by the stratosphere means that photolysis of O_2 cannot occur in the troposphere, and therefore that the Chapman cycle (see Section 3.1) cannot be initiated. Nevertheless, ozone is present and plays a crucial role in tropospheric photochemistry. In particular, the photolysis of tropospheric O_3 leads to formation of the OH radical. OH is the primary oxidant within the troposphere, and its abundance determines the ability of the atmosphere to cleanse itself of pollutants.

While tropospheric ozone can originate via transport from the stratosphere, this process is too slow to yield sufficient O_3 and hence OH to drive oxidation in the troposphere, and therefore an alternative mechanism is required for ozone formation. The dominant mechanism involves nitrogen oxides, NO_x, and the subsequent oxidation cycles initiated by OH. These cycles, which couple OH, HO_2, and peroxy (RO_2) radical species, remove primary emitted trace species that are harmful to humans (e.g. CO, benzene) or to the wider environment (e.g. greenhouse gases, GHGs, such as CH_4 and HFCs). However, many of the secondary products produced by photooxidation are also directly harmful. These include O_3, NO_2, and acidic and multifunctional species. Many of the latter have low volatility, and are able to partition effectively into the condensed phase to generate secondary organic aerosol (SOA). A famous example of the enhanced formation of such secondary pollutants is found in the occurrence of photochemical smog and the accelerated photooxidation chemistry behind this effect is also discussed.

4.1 Tropospheric Photochemistry of Ozone and the O(^{1}D) Radical

The atmosphere is an oxidising medium, and therefore chemical change within the troposphere often involves the conversion of a fully or partially reduced species of anthropogenic or natural origin to a more oxidised form. The most abundant oxidants are O_2 and O_3. However, these are relatively unreactive, and oxidation of trace gases instead proceeds via reactions with radical species. Almost without exception, the first step in the daytime oxidation of any tropospheric species is their reaction with the hydroxyl radical, OH. The dominant daytime source of OH is the reaction between water vapour and electronically-excited oxygen atoms, O(^{1}D), formed in the photolysis of ozone. Understanding the detailed photochemistry of O_3 at wavelengths in the near-UV is therefore of crucial importance in determining the *oxidising capacity* of the troposphere. In particular, a knowledge of the quantum yield, Φ, for O(^{1}D) production as a function of wavelength and temperature is vital to an understanding of the chemistry of the lower atmosphere. As shown in Figure 2.3, O_3 has two bands in its near-UV absorption spectrum. Excited molecules formed by absorption in the intense Hartley band (*ca.* 220–310 nm) fragment predominantly along two spin-conserving dissociation channels:

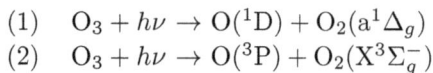

$$(1) \quad O_3 + h\nu \rightarrow O(^1D) + O_2(a^1\Delta_g)$$
$$(2) \quad O_3 + h\nu \rightarrow O(^3P) + O_2(X^3\Sigma_g^-)$$

Channel (1) dominates, with a quantum yield of *ca.* 0.9. At wavelengths longer than 310 nm, the Hartley band is overlapped by the weaker Huggins band, and it becomes energetically impossible to form the singlet products of Channel (1) for parent ozone molecules with zero internal energy. However, O_3 dissociation to form O(^{1}D) does take place at these longer wavelengths, as shown in Figure 4.1. This can occur both through the spin-forbidden reaction $O_3 + h\nu \rightarrow O(^1D) + O_2(X^3\Sigma_g^-)$ and by photolysis of vibrationally-excited ozone molecules. In the latter case, while only a small fraction of O_3 molecules possess enough internal energy to allow them to be dissociated in the spin-allowed reaction (1), the probability of excitation (quantified by the Franck–Condon factor) is large, and this channel can

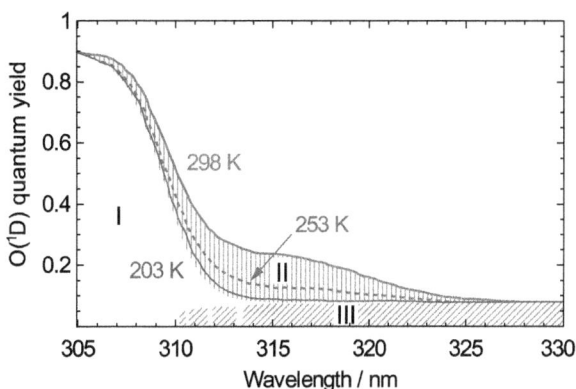

Figure 4.1 Recommendation values of $O(^1D)$ quantum yields from the photolysis of O_3 at 203, 253, and 298 K in the wavelength range 305–330 nm (solid lines). Contributions made by the various dissociation processes to the quantum yield are also indicated. Region I corresponds to $O(^1D)$ formation following excitation of parent vibrationless molecules and dissociation via the channel, $O(^1D) + O_2(a^1\Delta_g)$. Region II (vertical hatching) indicates the contribution from the hot-band excitation process leading to $O(^1D)$ formation via the channel, $O(^1D) + O_2(a^1\Delta_g)$, at 298 K, while region III (diagonal hatching) represents the contribution from the spin-forbidden process leading to $O(^1D)$ formation via the channel, $O(^1D) + O_2(X^3\Sigma_g^-)$.

Source: Adapted from Y. Matsumi *et al.*, *J. Geophys. Res.* **107** ACH 1-1, DOI: 10.1029/2001JD000510 (2002).

therefore play a significant role in atmospheric chemistry. The population of the vibrationally-excited levels is temperature dependent, and so the quantum yield of $O(^1D)$ production is also a function of temperature, as shown in Figure 4.1.

The reason why the data shown in Figure 4.1 are so important is that in the spectral region beyond 310 nm the actinic flux increases markedly, as shown in Figure 4.2. The rate of production of $O(^1D)$, and hence OH, also varies as a function of λ, also shown in Figure 4.2. In many situations, the formation rate of $O(^1D)$ at wavelengths longer than the 310 nm is highly significant; the photolysis rate constant has a maximum value of $\sim 3 \times 10^{-5}\,s^{-1}$,[1] small compared to the $\sim 2 \times 10^{-3}\,s^{-1}$ found in the stratosphere ($z = 40$ km), but the small amount of $O(^1D)$ produced nevertheless plays a crucial role in atmospheric chemistry, particularly at high latitudes from late autumn to early spring.

$O(^1D)$ has *ca.* $15000\,cm^{-1}$ of internal energy, and is therefore highly reactive. At high pressures, $O(^1D)$ undergoes rapid collisional quenching

---

[1]This is just the integrated area under the data displayed in Figure 4.2(b).

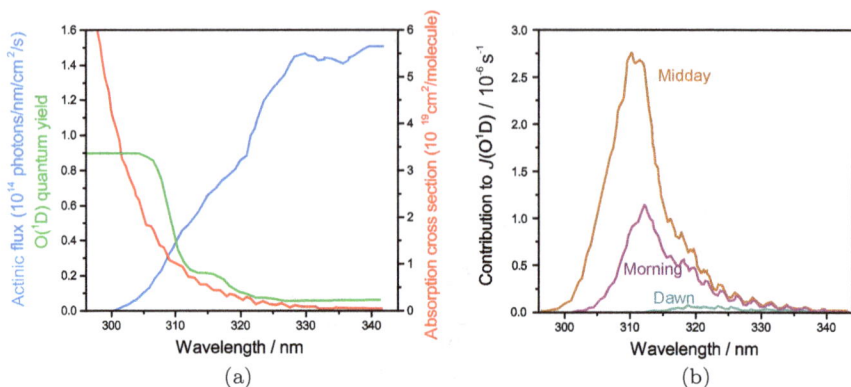

Figure 4.2   (a) The wavelength dependence of the absorption cross-section for O_3, the quantum yield for $O(^1D)$ production from the photolysis of O_3 at $298\,K$, and the actinic flux (SZA $\sim 15°$). (b) The wavelength dependence of the rate constant for production of $O(^1D)$ at different times of day i.e. at different values of the solar zenith angle.

Source: Adapted from A. Hofzumahaus *et al.*, *J. Geophys. Res.* **109** D08S90 (2004).

to the $O(^3P)$ electronic ground state, with a rate constant in air of $k_q \approx 3 \times 10^{-11}\,cm^3$ molecule$^{-1}\,s^{-1}$. However, in competition with electronic quenching, around 10% of the $O(^1D)$ reacts with water vapour to form OH via the reaction

$$O(^1D) + H_2O \rightarrow 2OH \qquad (4.1)$$

with the exact fraction of $O(^1D)$ lost by this pathway depending on the relative humidity. The reaction with H_2O is the dominant, but not the only, production pathway for OH radicals in the troposphere. In contrast, the reaction of $O(^3P)$ with H_2O is extremely endothermic, and so does not occur to any appreciable extent. As $O(^1D)$ is formed by photolysis, there is a strong diurnal variation in OH concentration, as shown in Figure 4.3. [OH] normally peaks at just a few million molecules per cm^3, reflecting the low source rate of OH and its fast sinks.[2] As we shall see in Section 4.2, reactions of OH are key to the 'self-cleansing' of the troposphere, with the main sinks

---

[2]Interestingly, Figure 4.3 shows an extremely high mean [OH] of *ca.* 1.5×10^7 cm^{-3} around noon, which persists despite the high reactivity of OH. Such unexplained high OH mixing ratios in the presence of high levels of biogenic VOCs and low to moderate levels of NO_x is a contemporary topic in atmospheric chemistry research. In contrast, measured OH levels under clean air conditions (low NO_x and low VOCs) and in highly polluted areas (high NO_x and high VOCs) can be explained reasonably well by current models.

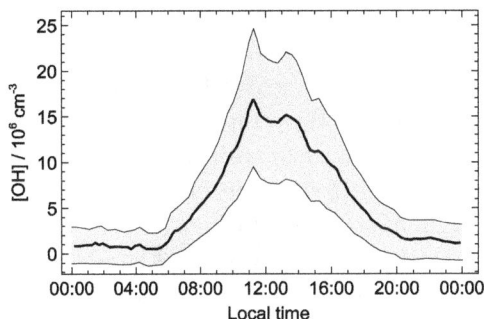

Figure 4.3 Diurnal variation of [OH] near Guangzhou, China, between 5 and 25 July 2006. The thick black line denotes the half-hourly mean diurnal profile and the enclosing thin grey lines represent the maximum data variability caused by the measurement instrument.

Source: Adapted from A. Hofzumahaus *et al.*, *Science* **324** 1702 (2009).

being reactions with CO and CH$_4$. In addition, OH concentrations are both spatially and temporally highly variable.

The steady-state approximation (SSA) can be applied to the concentration of O(^{1}D), as there is slow production due to low intensity of radiation of the required energy, and O(^{1}D) is immediately used up by a combination of rapid collisional quenching and chemical reaction. Considering only the primary quenching process gives

$$\frac{d[O(^1D)]}{dt} = J_{O(^1D)}[O_3] - k_q[O(^1D)][M] = 0 \qquad (4.2)$$

where $J_{O(^1D)}$ is the rate constant for production of O(^{1}D) via O$_3$ photolysis and k_q is the rate constant for collisional quenching of O(^{1}D) by atmospheric gases. When we solve this equation to find [O(^{1}D)], we find that at steady state, there is only around 2 g of O(^{1}D) within the troposphere.

4.2 The OH Radical: The Atmosphere's Detergent

For many species in the troposphere, photolysis is not an efficient sink, and their lifetimes τ are instead determined by reaction with the OH radical, e.g. OH oxidises SO$_2$, NO$_2$, CO, and volatile organic compounds (VOCs). The type of reaction that OH undergoes with a given hydrocarbon depends on the structure of the hydrocarbon. For saturated molecules, reaction is generally through abstraction of a hydrogen atom, while for unsaturated species reaction occurs through electrophilic addition across a double bond.

Of particular note, oxidation by OH is the only major loss pathway for tropospheric CH_4 and CO. In contrast, fully halogenated hydrocarbons do not react with OH, and hence chlorofluorocarbons (CFCs) persist long enough to be transported into the stratosphere, where they are photolysed. As discussed in Chapter 3, this can have catastrophic consequences for the stratospheric ozone layer. Nitrous oxide, N_2O, provides another example of a species which is immune to OH attack.

The OH radical cleanses the troposphere, thereby minimising pollution. However, we will begin our discussions by considering a *clean* environment, with no pollution and therefore no NO_x. In this case the chemistry of OH is dominated by reactions with CO and CH_4, which occur with (relative) branching ratio 0.7:0.3.

4.2.1 Reaction of OH with CO

CO is released into the troposphere through incomplete combustion and oxidation of CH_4 and other hydrocarbons and has a tropospheric mixing ratio of ~200 ppbv. Its reaction with OH occurs as follows and yields the hydroperoxy radical, HO_2 (steps (1) and (2)).

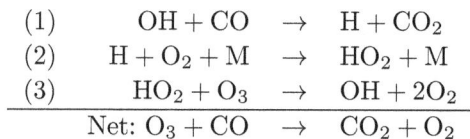

$$
\begin{array}{rrcl}
(1) & OH + CO & \rightarrow & H + CO_2 \\
(2) & H + O_2 + M & \rightarrow & HO_2 + M \\
(3) & HO_2 + O_3 & \rightarrow & OH + 2O_2 \\
\hline
& \text{Net: } O_3 + CO & \rightarrow & CO_2 + O_2
\end{array}
$$

In a clean environment, HO_2 can then react with O_3 to regenerate OH via reaction (3). The resulting reaction cycle establishes a HO_x ($=OH + HO_2$) equilibrium.[3] The second reaction in the cycle is very fast — it has an activation energy of zero as it is a radical–radical recombination, and in addition [M] is high — and therefore the first reaction, between OH + CO, is rate determining. The final reaction in the cycle regenerates OH *and* leads to ozone destruction. This means that overall, the photolysis of O_3, and the ensuing production of OH, is truly a sink for O_3. The overall oxidation mechanism for conversion of CO into CO_2 is shown in Figure 4.4(a). The principal removal pathway for HO_2 is its self-reaction to form the reservoir species H_2O_2, i.e. $2HO_2 \rightarrow H_2O_2 + O_2$. Because H_2O_2 is soluble in water, it

---

[3]See Section 3.2.2 for a discussion of a similar equilibrium within the stratosphere.

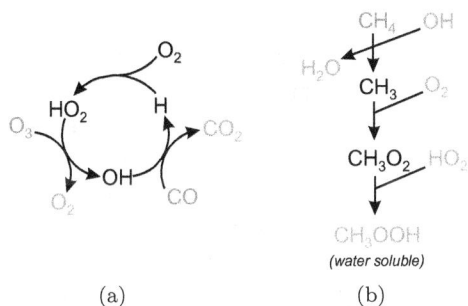

(a)   (b)

Figure 4.4   OH chemistry in a clean environment: (a) reaction with CO; (b) reaction with CH$_4$.

is mostly absorbed into clouds and rained out. H$_2$O$_2$ can also be photolysed back to HO$_x$, but this is a slow process.

The chemical lifetime of OH is less than 1 s, and the steady-state concentration of OH can be estimated as follows. The rate of OH production is

$$\frac{d[OH]}{dt} = 2f J_{O(^1D)}[O_3] = 0.2 J_{O(^1D)}[O_3] \tag{4.3}$$

where f ($\sim$0.1) is the fraction of O(^{1}D) which reacts with water to form OH. Assuming that the loss rate of OH is determined solely by reaction with CO, i.e. ignoring the contribution from reaction with CH$_4$ and other VOCs, the rate of OH loss is

$$\frac{d[OH]}{dt} = -k_1[OH][CO] \tag{4.4}$$

Applying the SSA to the OH concentration, and employing typical values for the various quantities ($J_{O(^1D)} \sim 10^{-5}\,\text{s}^{-1}$, $[O_3] \sim 20$ ppbv, $[CO] \sim 100$ ppbv, and $k_1 = 2 \times 10^{-13}\,\text{cm}^3$ molecule^{-1} s^{-1}), yields a value for [OH] of 2×10^6 molecules cm^{-3}, in reasonable agreement with the data shown in Figure 4.3. Despite the low abundance of OH, we emphasise again that this radical is of vital importance in tropospheric oxidation chemistry.

4.2.2 Reaction of OH with CH$_4$

Methane is widespread within the troposphere, with an abundance of around 1.8 ppmv, and arises from wetlands, rice paddies, landfills and fossil fuel combustion. Methane is only removed from the troposphere by the reaction OH + CH$_4$ → H$_2$O + CH$_3$, and also results in HO$_x$ production as discussed below. The relatively high pressure in the troposphere means that the CH$_3$ radical formed from the H atom abstraction instantly reacts with oxygen via the reaction CH$_3$ + O$_2$ + M → CH$_3$O$_2$ + M to form the methyl peroxy radical, CH$_3$O$_2$, with a rate constant of $\sim 1 \times 10^{-11}$ cm^3 molecule^{-1} s^{-1} at atmospheric pressure. The fate of CH$_3$O$_2$ depends on the chemical environment, i.e. whether it is HO$_2$ or NO$_x$ rich. In a clean environment, CH$_3$O$_2$ reacts with HO$_2$ to give methyl hydroperoxide: CH$_3$O$_2$ + HO$_2$ → O$_2$ + CH$_3$OOH. Again, this peroxide is water soluble and so is washed out, thereby acting as a sink for HO$_x$. As HO$_x$ is formed from O$_3$, the overall process leads to *loss* of O$_3$. The clean marine troposphere is therefore an O$_3$-destroying environment. The overall oxidation mechanism in which CH$_4$ is converted to CO$_2$ is shown in Figure 4.4(b). In summary, the oxidation of CO, CH$_4$ and other VOCs is initiated by OH and leads to ozone loss in unpolluted (low NO$_x$) conditions; OH is recycled in this process.

The chemical lifetime of HO$_2$ is between 5 and 100 s, and so the SSA can be applied to [HO$_2$] as well as to [OH]. As noted above, the main production route for HO$_2$ is via formation of OH radicals, and therefore the rate of change of [OH] is given by

$$\frac{d[OH]}{dt} = 2fJ_{O(^1D)}[O_3] + k_3[HO_2][O_3] - k_1[OH][CO] = 0 \qquad (4.5)$$

Again, only the rate of OH loss by the reaction OH + CO → H + CO$_2$ needs to be considered, as the reaction H + O$_2$ is so fast. The main loss route for HO$_2$ is 2HO$_2$ → H$_2$O$_2$ + O$_2$, which occurs with a rate constant k_4, and so

$$\frac{d[HO_2]}{dt} = -k_3[HO_2][O_3] + k_1[OH]\ [CO] - 2k_4[HO_2]^2 = 0 \qquad (4.6)$$

Finally,

$$\frac{d[HO_x]}{dt} = 2fJ_{O(^1D)}[O_3] - 2k_4[HO_2]^2 = 0 \qquad (4.7)$$

Rearranging the previous equations yields

$$[HO_2] = (fJ_{O(^1D)}[O_3]/k_4)^{1/2} \qquad (4.8)$$

$$[OH] = \frac{2fJ_{O(^1D)}[O_3] + k_3[HO_2][O_3]}{k_4[CO]} \qquad (4.9)$$

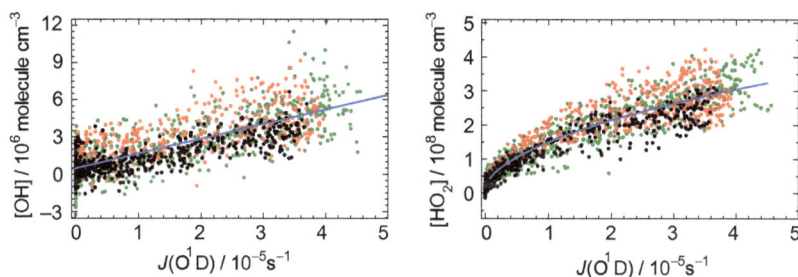

Figure 4.5 Measurements of [OH] and [HO$_2$] in the remote marine tropospheric boundary layer as a function of $J_{O(^1D)}$ at Cape Verde in 2009. Black dots denote winter (February) measurements while red and green represent summer (June and September) observations.

Source: Adapted from S. Vaughan *et al.*, *Atmos. Chem. Phys.* **12** 2149 (2012). Original figure kindly provided by D.E. Heard (Leeds).

The implications of these results are two-fold. Firstly, [HO$_2$] $\propto$ $(J_{O(^1D)})^{1/2}$, and secondly, the ratio of HO$_2$ to OH is *ca.* 100:1, so that HO$_2$ is by far the dominant component of HO$_x$. These conclusions are confirmed in Figures 4.5(a) and 4.5(b) which show measured HO$_2$ and OH concentrations as a function of the rate constant for O(^{1}D) production. Taking typical values of [O$_3$] = 25 ppbv, [CO] = 100 ppbv, and substituting into the equations above yields [HO$_2$] $\sim 10^8$ molecules cm^{-3} and [OH] $\sim 10^6$ molecules cm^{-3}. This [HO$_2$] value is in very good agreement with total peroxy concentration [HO$_2$] + [RO$_2$] (see later). It should also be noted that there is no HO$_x$ production at night because no photolysis occurs.

4.3 Field Measurements of OH and HO$_2$

The presence of OH and HO$_2$ is confirmed by field measurements, most commonly employing the method of fluorescence assay by gas expansion (FAGE).[4] In the FAGE technique, OH radicals are measured by UV laser-induced fluorescence (LIF) spectroscopy at low pressure (*ca.* 1 Torr). HO$_2$ is detected by first converting it into OH by the addition of NO, and then detecting the resultant OH fluorescence. The FAGE method utilises the (0,0) band of the A–X ($^2\Sigma^+ - ^2\Pi$) electronic transition around 308 nm, and the laser excitation and the fluorescence wavelengths are the same, i.e. resonant. The (0,0) spectral band is chosen due to various complicating

---

[4]See for example D. Stone, L.K. Whalley, and D.E. Heard, *Chem. Soc. Rev.* **41** 6348 (2012).

factors associated with detection via alternative bands such as the (1,0) or (0,1) bands. For example, laser excitation at *ca.* 280 nm on the (1,0) band leads to the photolysis of ambient O_3 which then leads to further OH production, yielding artificially high measured [OH], while emission on the (0,1) band is not sufficiently intense to provide the required detection sensitivity. When using resonance fluorescence, it is important to discriminate the resonance fluorescence from the more intense scattered light from the laser. This is achieved by employing delayed photon-counting detection to measure the fluorescence after the scattered laser pulse has cleared the detection region. Low pressure is required in order to reduce the competition between fluorescence and collisional quenching of the $^2\Sigma^+$ state. A typical detection limit for the FAGE technique is *ca.* $(2\text{--}5) \times 10^5$ and $(5\text{--}10) \times 10^5$ molecules cm^{-3} for OH and HO_2, respectively, depending upon the individual instrument and averaging period.

The OH radical may also be detected using chemical ionisation mass spectrometry (CIMS). In this technique, OH is converted quantitatively to $H_2^{34}SO_4$ by the reactions with $^{34}SO_2/O_2/H_2O$ and is subsequently chemically ionised by the reaction $NO_3^- \cdot HNO_3 + H_2^{34}SO_4 \rightarrow H^{34}SO_4^- \cdot HNO_3 + HNO_3$. The isotopically labelled ^{34}S allows discrimination against naturally occurring $H_2^{32}SO_4$. Finally, following collisional fragmentation of $H^{34}SO_4^- \cdot HNO_3$ the ratio $H^{34}SO_4^-/NO_3^-$ is measured in a quadrupole mass spectrometer. The CIMS method is the most sensitive of all OH field instruments, with a detection limit of better than 10^5 molecules cm^{-3}.[5]

The low concentrations of peroxy radicals HO_2 and RO_2, along with their high reactivity and short lifetimes, make these species difficult to measure. The only *direct* measurements of HO_2 and some organic RO_2 species have been made by matrix-isolation electron spin resonance (MIESR), but this method suffered from poor time resolution (several minutes), and is no longer used. As mentioned above, HO_2 may be measured indirectly using FAGE, following conversion to OH via reaction with NO. It should be noted that other peroxy radicals also react with NO to generate OH, so that in fact the FAGE method detects HO_2 and the sum of organic peroxy radicals. CIMS can also be used to detect peroxy radicals, but as the conversion chemistry takes place at atmospheric pressure, it is difficult to prevent the simultaneous conversion of RO_2 to OH. For these reasons it is typically the sum of peroxy radicals that is measured, although through

---

[5]D.E. Heard and M.J. Pilling, *Chem. Rev.* **103** 5163 (2003).

judicious control of the reagent gas flows, it is possible to measure HO_2 separately.

4.4 Nitrogen Oxides and the Production of Tropospheric Ozone

In Section 4.2, the oxidation chemistry of the clean troposphere was considered, and it was concluded that such an environment leads to the overall destruction of ozone. In this section, focus turns to reactions involving nitrogen oxides, NO_x, which instead lead to the formation of ozone. Anthropogenic sources of NO_x include fossil fuel combustion associated with energy and transport, biomass burning and aviation. Natural sources include lightning (mostly in the tropics) and emission from soils; this latter source is also affected by anthropogenic activities. The corresponding sinks are the wet deposition of nitrate over land and ocean, as well as dry deposition of NO_x. NO_x has a short lifetime, as it is rapidly converted to and then deposited as HNO_3, and therefore the highest $[NO_x]$ is found close to its source. It should however be noted that NO_x can be transported throughout the troposphere when converted to reservoir species such as peroxyacetyl nitrate (PAN) (Section 4.6), and so local changes can influence regional compositions.

We begin by considering a hypothetical atmosphere in which there are no peroxy radicals, and only consider NO, NO_2, and O_3. The relevant reactions are:

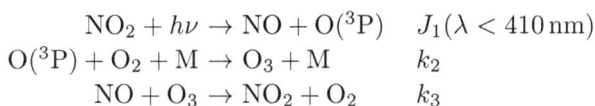

$$NO_2 + h\nu \rightarrow NO + O(^3P) \quad J_1(\lambda < 410\,\text{nm})$$
$$O(^3P) + O_2 + M \rightarrow O_3 + M \quad k_2$$
$$NO + O_3 \rightarrow NO_2 + O_2 \quad k_3$$

This is a null cycle for O_3, and establishes a steady-state concentration for this species. Applying the SSA to these three reactions yields the *Leighton equation*:

$$[O_3] = J_1 \frac{[NO_2]}{k_3[NO]} \tag{4.10}$$

with the same result being obtained if the SSA is applied to NO_2 and NO. The Leighton equation quantifies the day-time dependence of $[O_3]$ upon the relative amounts of the two members of the NO_x family. The value of J_1 corresponds to a lifetime of NO_2 of around $100\,\text{s}$ in the troposphere.

For typical tropospheric conditions, $[O_3] \sim 40$ ppbv, and $[NO_2] \sim 2[NO]$ during the day, while $[NO] = 0$ at night when no photolysis occurs.

The presence of NO_x dramatically changes the CO, CH_4, and general hydrocarbon oxidation chemistry discussed in Section 4.2, and leads to O_3 *production*. In the real atmosphere, the Leighton ratio is perturbed by the presence of other oxidants, mostly peroxy radicals, leading to the conversion of NO into NO_2 and the production of O_3. Two sets of reactions are important in this context, again pertaining to the oxidation of CO and hydrocarbons respectively:

$$\begin{aligned}
\textbf{Set 1} &\text{ — reaction of NO with } HO_2 \\
OH + CO &\rightarrow H + CO_2 \\
H + O_2 + M &\rightarrow HO_2 + M \\
NO + HO_2 &\rightarrow OH + NO_2 \\
NO_2 + h\nu &\rightarrow NO + O(^3P) \\
\underline{O(^3P) + O_2 + M} &\rightarrow \underline{O_3 + M} \\
\text{Net: } CO + 2O_2 + h\nu &\rightarrow CO_2 + O_3
\end{aligned}$$

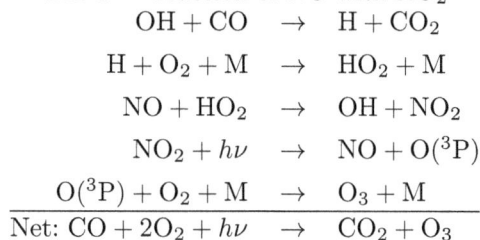

$$\begin{aligned}
\textbf{Set 2} &\text{ — reaction of NO with } RO_2 \\
OH + RH &\rightarrow H_2O + R \\
R + O_2 + M &\rightarrow RO_2 + M \\
NO + RO_2 &\rightarrow RO + NO_2 \\
RO + O_2 &\rightarrow R'CHO + HO_2 \\
\underline{NO + HO_2} &\rightarrow \underline{OH + NO_2} \\
\text{Net: } RH + 2NO + 2O_2 &\rightarrow R'CHO + H_2O + 2NO_2 \\
\text{e.g. } CH_4 + 2NO + 2O_2 &\rightarrow HCHO + H_2O + 2NO_2
\end{aligned}$$

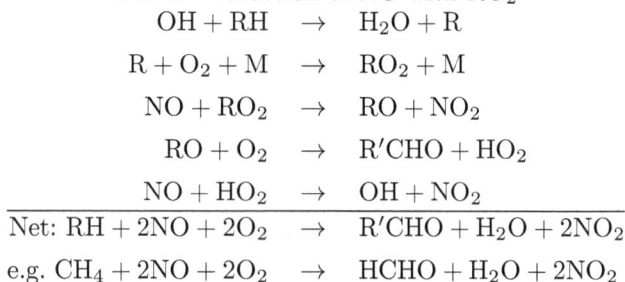

There are two key points here. Firstly, *both* OH and NO are regenerated in the cycle, i.e. it is catalytic with respect to HO_x and NO_x. Secondly, the photolysis of NO_2 leads to the production of ozone. The cycles are depicted in Figure 4.6. In the case of methane oxidation, the cycle results in the production of two ozone molecules, while for higher hydrocarbons, RH, even more O_3 is produced per molecule of RH oxidised. Although these reactions are slower than the $NO + O_3$ reaction within the Leighton scheme, they significantly perturb the *photostationary state*. This is because the reaction of NO with peroxy radicals leads to the oxidation of NO to

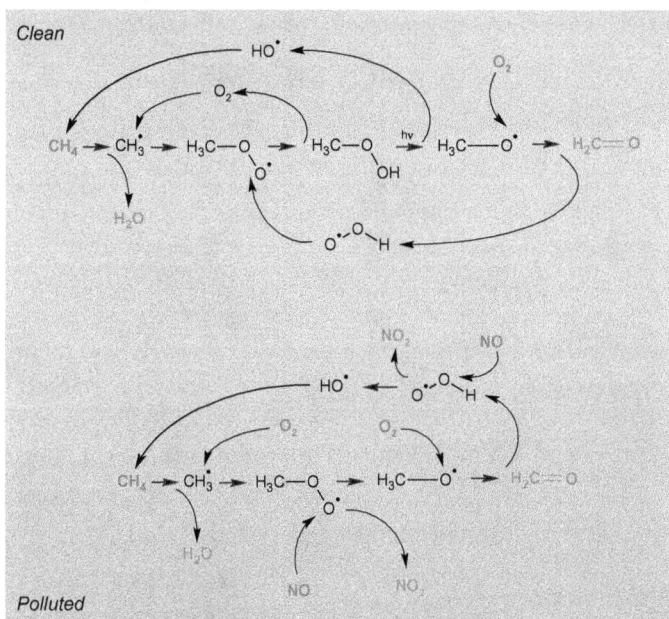

Figure 4.6   Comparison of CH_4 oxidation pathways in clean and NO_x rich (polluted) environments.

NO_2 *without* destroying O_3. The sequence leads to O_3 production when NO_2 is photolysed to give $O(^3P)$, which then forms O_3, i.e. every time NO_2 is produced, O_3 is produced. The rate of ozone production is

$$\frac{d[O_3]}{dt} = k_a[NO][HO_2] + k_b[NO][CH_3O_2] \tag{4.11}$$

where k_a and k_b are the rate constants for the reaction of NO with HO_2 and CH_3O_2 respectively. It should also be noted that the reaction $RO_2 + NO$ may also yield the alkyl nitrate product, $RONO_2$, with the yield of this product increasing as RO_2 increases in molecular size. Similarly, the alkoxy radicals, RO, not only react with O_2 but may also decompose and/or undergo intramolecular isomerisation; where isomerisation is possible it dominates, and is increasingly prevalent as R increases in size.

One of the photo-oxidation products is formaldehyde, HCHO. This is the most abundant and important organic carbonyl compound in the atmosphere, with levels ranging from ~50 pptv in clean tropospheric air to ≤70 ppbv in urban areas. Interestingly, HCHO undergoes photolysis via two pathways, with threshold wavelengths as indicated[6]:

$$HCHO + h\nu \rightarrow HCO + H \quad (\lambda \leq 330\,nm)$$
$$HCHO + h\nu \rightarrow H_2 + CO \quad (\lambda \leq 361\,nm)$$

The first pathway is referred to as the *radical channel*, whereas the second pathway is referred to as the *molecular channel*. The products of both pathways play a significant role in atmospheric chemistry. The CO produced by the molecular channel undergoes further oxidation to produce CO_2, while the H atom and formyl radical that are produced by the radical channel undergo rapid reactions with O_2 to produce HO_2. The production of two molecules of HO_2 acts as a positive feedback to accelerate the rate of hydrocarbon oxidation. The reaction $RH + NO_x$ therefore has the dual effect of increasing both ozone *and* HO_2 production, which speeds up the chemistry.

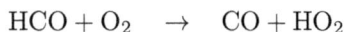

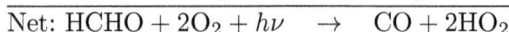

$$
\begin{array}{rcl}
HCHO + h\nu & \rightarrow & HCO + H \\
HCO + O_2 & \rightarrow & CO + HO_2 \\
H + O_2 + M & \rightarrow & HO_2 + M \\
\hline
\text{Net: } HCHO + 2O_2 + h\nu & \rightarrow & CO + 2HO_2
\end{array}
$$

4.5 Oxidation Chemistry at Night

Radical chemistry in the polluted night-time troposphere is governed by the abundance of NO_3 radicals, which readily oxidise alkenes, aldehydes, and biogenic VOCs. NO_3 plays a negligible role in day-time oxidation chemistry due to its efficient photolysis by visible radiation, but at night its importance is comparable to that of OH radicals during daytime. The key reactions controlling the formation and destruction of *oxidised nitrogen* are shown in Figure 4.7, and can be summarised as follows. Tropospheric NO_3 radicals are formed through the reaction of nitrogen dioxide with ozone: $NO_2 + O_3 \rightarrow NO_3 + O_2$. NO_3 radicals do not build-up to relevant levels (<1 pptv) during the day because they efficiently absorb light in the visible region of the solar spectrum, leading to photolysis into a radical channel, $NO_3 + h\nu \rightarrow NO_2 + O$, and a molecular channel, $NO_3 + h\nu \rightarrow NO + O_2$.[7]

---

[6]See for example F.D. Pope *et al.*, *Faraday Discuss.* **130** 59 (2005).
[7]The radical channel dominates for $\lambda < 590\,nm$, while the molecular channel prevails above 615 nm, although it should be noted that NO_3 fluorescence is the dominant decay process following excitation at wavelengths above 605 nm.

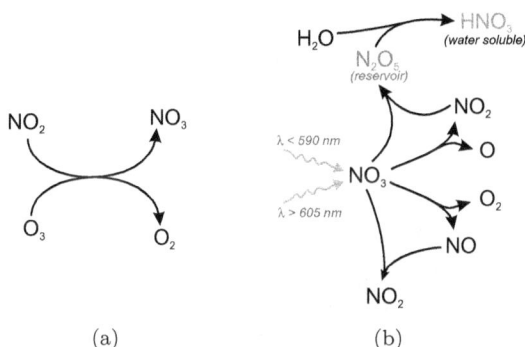

Figure 4.7   A schematic of NO$_3$ (a) formation and (b) loss routes.

The lifetime of NO$_3$ under typical daylight conditions is *ca.* 5 s. In addition, NO$_3$ reacts very rapidly with NO via the reaction NO$_3$ + NO → 2NO$_2$. This reaction is important during the day, particularly in cases of fresh emissions of NO, when its rate often exceeds that of photolysis. At night, the reaction with NO often limits the lifetime of NO$_3$ when no excess O$_3$ is present to convert NO into NO$_2$. NO$_3$ itself reacts with NO$_2$ to form N$_2$O$_5$. N$_2$O$_5$ is thermally unstable and readily decomposes into its precursors, thereby establishing an equilibrium between NO$_3$ and N$_2$O$_5$. The equilibrium is strongly temperature dependent, with N$_2$O$_5$ dominating at low temperatures. N$_2$O$_5$ can be efficiently hydrolysed to HNO$_3$ on the surface of aerosol particles: N$_2$O$_5$ + H$_2$O (surface) → 2HNO$_3$, and the wet and dry deposition of the HNO$_3$ formed by hydrolysis of N$_2$O$_5$ is one of the most important loss reactions of oxidised nitrogen compounds in the atmosphere.

NO$_3$ is abundant during the night when OH cannot form, and in addition, can undergo oxidation reactions which OH cannot. For example, NO$_3$ competes with OH to react with unsaturated VOCs such as isoprene (the most prevalent biogenic VOC) and terpenes; in the latter case there are many double bonds to react with, and a correspondingly large number of secondary products that are able to partition effectively to the condensed phase, creating SOA (see Section 4.7). A schematic of NO$_3$-driven VOC oxidation is shown in Figure 4.8. NO$_3$ can add across either side of a double bond, and subsequent addition of O$_2$ leads to the formation of distinct peroxy radicals, which then react further to produce aldehydes, organic nitrates, NO$_2$, and HO$_2$. The branching ratio between NO$_2$ and HO$_2$ production is governed by the side of the double bond to which NO$_3$

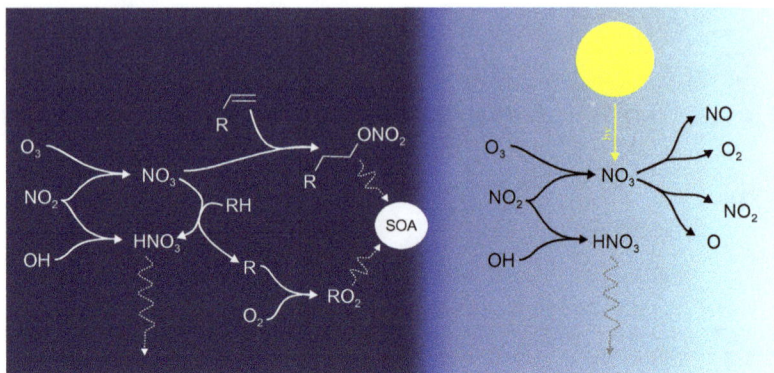

Figure 4.8   Comparison of tropospheric reactions involving the nitrate radical at night (left) and during the day (right).

adds, which in turn depends on the structure of the organic molecule. It should also be noted that in contrast to OH, NO_3 is lost during oxidation, whereas OH is recycled from the reaction $HO_2 + NO$.

The mixing ratio of NO_3 radicals is typically in the low pptv range, at least an order of magnitude larger than day-time [OH] values, and its spatial and temporal variability can be high. This places high demands on the selectivity, sensitivity, and time resolution of measurement techniques used for NO_3 detection. High-quality, accurate, and precise *in-situ* measurements of NO_3 and N_2O_5 are fundamental for understanding the processes controlling the chemistry of nitrogen oxides and their role in determining the oxidising capacity of the night-time troposphere. A variety of methods have been employed to measure NO_3 in the field; these include long-path differential optical absorption spectroscopy (DOAS), MIESR, LIF, and cavity enhanced absorption spectroscopies.[8] Some of these instruments are also capable of measuring $[N_2O_5]$ via quantitative thermal conversion into NO_3 in a heated detection cell, with $[N_2O_5]$ being obtained after subtraction of the (generally much smaller) $[NO_3]$ signal. Figure 4.9 shows an example of field measurements of NO_3 and N_2O_5 performed via broadband cavity enhanced absorption spectroscopy (BB-CEAS), in which the broadband cavity transmission is dispersed onto a CCD array. In

---

[8]See, for example, *Cavity RingDown Spectroscopy: Techniques and Applications*, eds. G. Berden and R. Engeln, Wiley (2009) and Chapter 6 of *Astrochemistry: From the Big Bang to the Present Day* by C. Vallance.

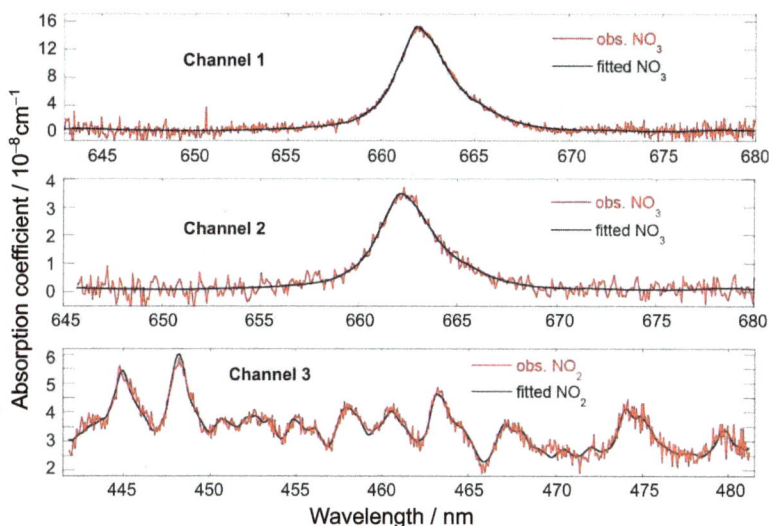

Figure 4.9   Exemplar field measurements of a selection of nitrogen oxides by BB-CEAS.
Source: Adapted from O.J. Kennedy *et al.*, *Atmos. Meas. Tech.* **4** 1759 (2011). The channels measure ambient and dissociated (from N_2O_5) NO_3 (548±3 pptv), only ambient NO_3 (80 ± 1 pptv), and NO_2 (3.95 ± 0.01 ppbv) respectively. Original figure kindly provided by S.M. Ball (Leicester).

contrast to cavity ring-down spectroscopy (CRDS), in which the temporal decay of light intensity exiting the cavity is measured, here it is the time-integrated light intensity transmitted by the cavity with, I, and without, I_0, the sample, that is determined. I and I_0 are related to the absorption coefficient of the sample (per unit path length), $\kappa(\lambda)$, by the relationship

$$\frac{(I_0 - I)}{I} = \kappa(\lambda)\frac{l}{(1 - R(\lambda))} \tag{4.12}$$

where $R(\lambda)$ is the average reflectivity of the cavity mirrors.

While NO_3 is the dominant oxidant at night, it should be noted that O_3 also plays a role, particularly in the case of unsaturated species. Though the rate constants for ozonolysis are orders of magnitude smaller than those for reaction with NO_3 or OH, ozonolysis is still important because of the relatively large abundance of O_3, i.e. [OH] $\sim$ 0.1 pptv and $[O_3]$ $\sim$ 100 ppbv. The gas-phase ozonolysis of alkenes is an important sink for both O_3 and alkenes, and has been recognized as a main source of organic acids and

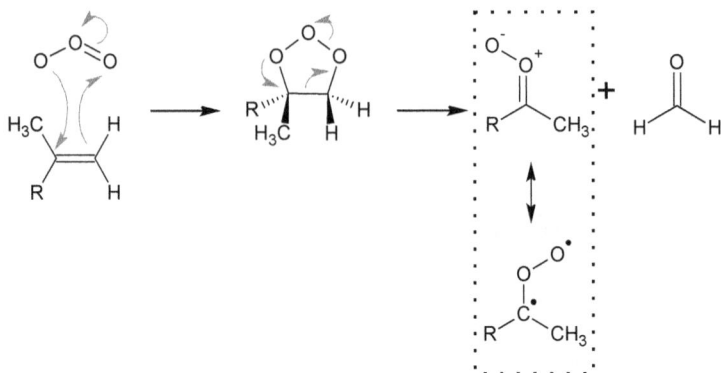

Figure 4.10   Formation of CIs (dotted box) in the ozonolysis of alkenes.

hydroperoxides in the atmosphere. The gas-phase ozonolysis of biogenic alkenes, typically terpenes, is also known to form particulate products which serve as the precursors to SOA. Ozonolysis proceeds as shown in Figure 4.10, and involves the formation of internally excited Criegee radicals which may be collisionally stabilised to form *Criegee intermediates* (CIs) or undergo decomposition and/or isomerisation.

CIs have been detected in laboratory experiments and predicted by theory to oxidise SO_2, NO_2, and other trace gases. Furthermore, recent field studies in a boreal forest[9] and at a coastal site[10] have identified a missing process in the oxidation of SO_2 to H_2SO_4 and have inferred CIs as the possible oxidant. Interestingly, both experimental and theoretical studies have shown that the reaction kinetics of CIs are highly structure-dependent; for example, for reaction with water vapour, *syn*-CIs, in which the alkyl group is on the same side as the terminal oxygen of the carbonyl oxide group, react very slowly, whereas *anti*-CIs react relatively fast.[11] A potentially important fate of CIs is unimolecular decomposition and this is therefore likely to be a significant sink for *syn*-CIs due to their slow reaction with water vapour. The hydroperoxide rearrangement has been identified as being dominant for *syn*-CIs, and acts as a non-photolytic source of atmospheric oxidants. By comparison, *anti*-CIs primarily undergo

---

[9]R.L. Mauldin III *et al.*, *Nature* **488** 193 (2012).

[10]H. Berresheim *et al.*, *Atmos. Chem. Phys.* **14** 12209 (2014).

[11]M.J. Newland *et al.*, *Phys. Chem. Chem. Phys.* **17** 4076 (2015) and *Atmos. Chem. Phys.* **15** 9521 (2015).

rearrangement and decomposition via a dioxirane intermediate, producing a range of products that contribute to the overall HO_x yield.

4.6 NO_x Reservoirs and Transporters: PAN

It is clear from Sections 4.2 and 4.4 that at a certain concentration of NO, the atmosphere will change from being O_3 destroying to being an O_3-producing environment. Net O_3 production will occur when [NO] is high enough for the reaction $NO + HO_2$ (which produces NO_2 which is rapidly photolysed to give $O(^3P)$ which then generates ozone) to compete effectively with the reaction $O_3 + HO_2$ (which destroys ozone). Consider the situation in which the ozone production and destruction terms are balanced, i.e. $k_a[NO][HO_2] = k_b[O_3][HO_2]$. Taking values of the rate constants k_a and k_b to be 7.7×10^{-12} and 1.9×10^{-15} cm^3 molecule^{-1} s^{-1}, respectively, at 298 K, leads to the conclusion that $[NO]/[O_3] \sim 2.5 \times 10^{-4}$, i.e. if $[O_3]$ is typically 40 ppbv then [NO] will only be 10 pptv. However, typical urban environments have vastly greater amount of NO_x ($\sim$30 ppbv) and so are definitely O_3-producing environments.

It is also clear that any mechanism that can either remove NO_x from the atmosphere or cause it to be transported away will impact on O_3 levels, especially as so little NO is needed for net O_3 production. The dominant sink for NO_x is the formation of nitric acid: $OH + NO_2 + M \rightarrow HNO_3 + M$, which is also a sink for OH. This daytime reservoir for NO_x and HO_x is readily lost by wet and dry deposition and by aerosol uptake. At high $[NO_x]$, reaction between OH and NO_2 controls the net loss of HO_x such that there is a decrease in the steady-state value of $[HO_x]$. At night, NO_3 is the primary oxidant and formation of N_2O_5 is the sink as it undergoes hydrolysis as detailed in Section 4.5.

Peroxyacetyl nitrate (PAN), $CH_3COO_2NO_2$, is an effective NO_x reservoir formed in polluted environments by the degradation of higher hydrocarbons in the presence of NO_2; this is shown schematically in Figure 4.11(a). The formation of PAN is reversible: $CH_3C(O)O_2 + NO_2 \rightleftharpoons CH_3C(O)O_2NO_2$, with forward and backward rate constants k_1 and k_{-1}, respectively. In addition, peroxyacetyl radicals undergo a competitive reaction with NO, $CH_3C(O)O_2 + NO \rightarrow CH_3 + CO_2 + NO_2$, with rate constant k_2; the CH_3 product is then oxidised to HCHO by reactions outlined earlier. PAN is insoluble in water and so is not susceptible to rain or wash out. In addition, other sinks for PAN, such as reaction with OH and photolysis, are inefficient, and so [PAN] can be as high as 70 ppbv.

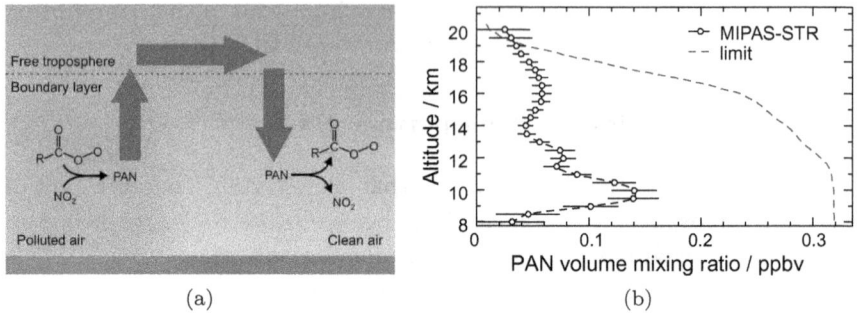

(a)   (b)

Figure 4.11   (a) PAN formation and transport; (b) PAN vertical profile obtained from measurements by the Michelson Interferometer for Passive Atmosphere Sounding-STRatospheric aircraft (MIPAS-STR). Also shown is the upper limit for the abundance of PAN corresponding to the simultaneously measured volume mixing ratio of NO_y (HNO_3, $ClONO_2$, HO_2NO_2, PAN).

Source: Adapted from C. Keim *et al.*, *Atmos. Chem. Phys.* **8** 4891 (2008).

Applying the SSA to the peroxyacetyl radicals involved in the three reaction above allows the lifetime of PAN to be expressed as

$$\frac{d[PAN]}{dt} = -k_{-1}\left(1 - \frac{1}{1 + \frac{k_2[NO]}{k_1[NO_2]}}\right)[PAN] = -(1/\tau)[PAN] \qquad (4.13)$$

The rate of decomposition of PAN depends on temperature and the relative amounts of NO and NO_2 present. For example, PAN has a longer lifetime at night, as $[NO] = 0$. The lifetime τ of PAN is *ca.* 1 hour at 295 K in the lower troposphere, but increases to several months at 250 K, a typical upper tropospheric temperature. Above the planetary boundary layer (PBL), PAN is stable — see Figure 4.11(b) — and can transport significant amounts of NO_x to clean areas of the atmosphere. Higher temperatures and low local abundances of NO_2 result in PAN dissociating to release NO_x, leading to O_3 production at distances far removed from pollution sources.

4.7 Photochemical Smog

As discussed in Section 4.4, sunlight, NO_x, and VOCs combine to form O_3, which is toxic to humans and vegetation because it oxidises biological tissue. Ozone generation is a feature of most conurbations, especially in summer, as most famously observed in the Los Angeles basin and illustrated in Figure 4.12. *Photochemical smog* (the latter word is a contraction of smoke and fog) is readily discerned, and pertains to a situation in which

(a)

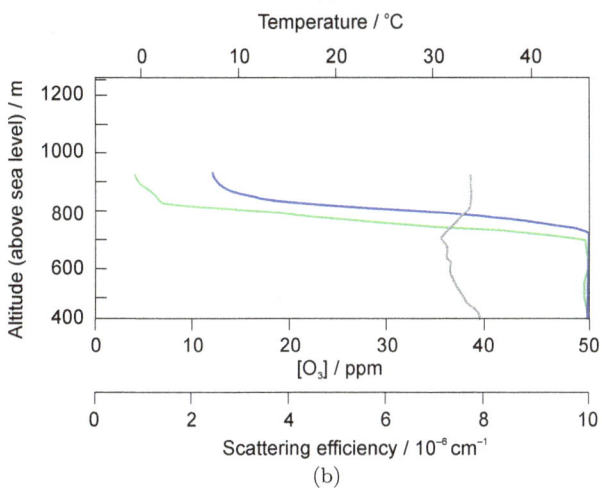

(b)

Figure 4.12   Photochemical smog in Los Angeles. The upper panel (a) shows sunset on a smoggy summer day in Los Angeles during the early 2000s. The lower panel (b) shows some typical physicochemical parameters associated with photochemical smog; green is the particle scattering efficiency, blue is the ozone concentration, and grey is the temperature.

Source: (a) Photograph by Jon Sullivan and available from http://www.public-domain-photos.com/travel/los-angeles/los-angeles-smog-2-free-stock-photo-4.htm.
(b) Adapted from D.L. Blumenthal *et al.*, *Atmos. Environ.* **12** 893 (1978).

visibility is significantly impaired as a result of local air pollution. The loss of visibility is generally caused by the formation of fine particles, which are themselves a health hazard. The presence of these particles is the final result of substantial enhancement in the natural process of cleansing the atmosphere of VOCs through OH oxidation in the presence of high NO_x, coupled with pollution being trapped close to the ground by a temperature inversion which acts as a lid over the basin. During summer, the inversion layer extends some 1 km above the surface. In heavily polluted air, a wide variety of VOCs are available to react with OH, and these reactions occur faster than reactions of OH with CH_4 and CO. The result is that interconversion rates between OH and RO_2 and between NO and NO_2 are higher, leading to increased O_3 production.

We again emphasise the following oxidation cycle:

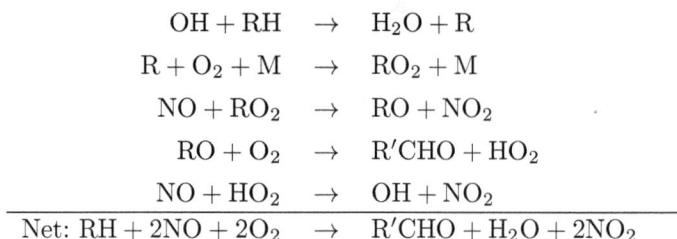

$$
\begin{array}{rcl}
OH + RH & \to & H_2O + R \\
R + O_2 + M & \to & RO_2 + M \\
NO + RO_2 & \to & RO + NO_2 \\
RO + O_2 & \to & R'CHO + HO_2 \\
NO + HO_2 & \to & OH + NO_2 \\
\hline
\text{Net: } RH + 2NO + 2O_2 & \to & R'CHO + H_2O + 2NO_2
\end{array}
$$

R'CHO is however not the end of the cycle, as it can then react with OH, leading to further HO_x and O_3 production, e.g. many VOCs can generate five to ten O_3 molecules (per VOC molecule oxidised). It should also be noted that rather than termination by formation of the peroxides H_2O_2 or ROOH, i.e. via loss of HO_x, termination in polluted air occurs by the formation of HNO_3, either directly or via N_2O_5. Furthermore, an increase in the surface temperature leads to faster reactions and higher biogenic VOC emissions, both of which act as a positive feedback to enhance production of O_3. Finally, it should be noted that oxygenated VOCs (O-VOCs) such as aldehydes deposit rapidly onto existing aerosol and so change their size and composition; we return to this point later.

We now consider the net rate of ozone production, $R(O_3)$, as a function of $[NO_x]$, focussing upon ozone destruction and production routes, and the loss routes for NO_x. O_3 destruction is via reaction with HO_2, and the rate is independent of $[NO_x]$. The rate of O_3 production, however, increases with increasing $[NO_x]$, as does the dominant loss process for NO_x, formation of HNO_3. Figure 4.13 shows a schematic of $R(O_3)$ as a function of $[NO_x]$,

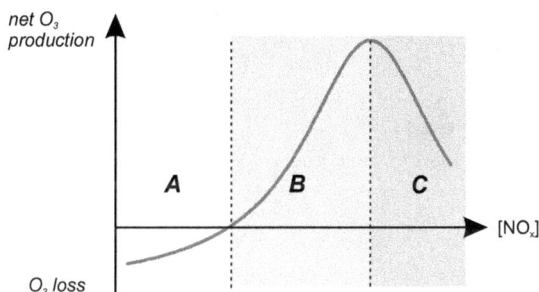

Figure 4.13   A schematic of O_3 production and loss rates as a function of NO_x. Adapted from S.C. Liu and M. Trainer, *J. Atmos. Chem.* **6** 221 (1988).

in which three regions, denoted A, B, and C, are discernible. Region A corresponds to clean troposphere with sufficiently low $[NO_x]$ that it is net O_3 destroying i.e. $R(O_3) < 0$. $R(O_3)$ increases with increasing NO_x, and eventually, the rate of O_3 production equals the rate of O_3 destruction, and so $R(O_3) = 0$. Region B corresponds to the polluted troposphere with high $[NO_x]$, and is net O_3 producing, $R(O_3) > 0$. Now the rate of O_3 production is greater than rate of O_3 destruction, and continues to increase with increasing $[NO_x]$, until reaching a maximum value which defines the start of Region C. In this region (C), $R(O_3)$ decreases with increasing $[NO_x]$; this is because increasing $[NO_x]$ also increases the rate of formation of the HNO_3 reservoir by reaction of OH with NO_2. The increased rate of this termination reaction means that the free-radical catalysed production of O_3 is less efficient, i.e. the formation of reservoirs inhibits VOC oxidation and hence O_3 production.

In general, there is a nonlinear relationship between O_3 production and the initial concentration of pollutants (VOCs + NO_x). This relationship is commonly displayed in the form of an *isopleth diagram*. An isopleth is a contour line of a function of two variables, and is generally a curve along which the function has a constant value. An exemplar isopleth diagram is shown in Figure 4.14; it can be seen that Figure 4.13 corresponds to a cross-section of the isopleth at a fixed value of [VOC]. Interestingly, there are two limiting regimes for ozone production denoted on the isopleth, termed *NO_x-limited* and *VOC-limited* respectively. In the former case there is net O_3 production in moderate/high $[NO_x]$ environments and the O_3 production may be decreased by decreasing $[NO_x]$. This limit is appropriate for many tropical and Southern Hemisphere locations. As noted above, very high $[NO_x]$ inhibits O_3 production, as increased rates of termination

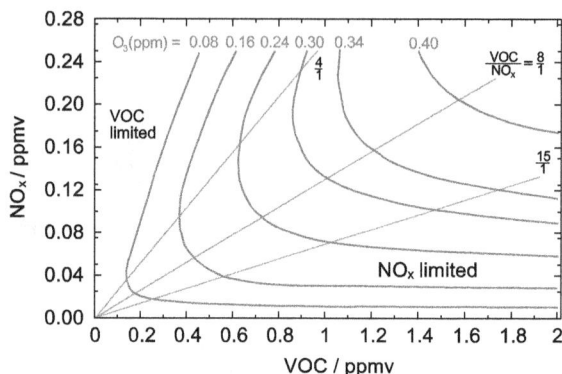

Figure 4.14   An example of an isopleth diagram showing VOC and NO_x-limited O_3 production.

Source: Adapted from *Rethinking the Ozone Problem in Urban and Regional Air Pollution*, National Academy Press (1991).

reactions cause radical concentrations to fall. In the VOC-limited region, the strategy for decreasing $[O_3]$ is to decrease [VOC] rather than (as one might suspect) to decrease $[NO_x]$. In this limit, a decrease in $[NO_x]$ will actually increase the abundance of O_3. When developing strategies to combat photochemical smog, it is therefore most effective to establish tighter emission restrictions on the more reactive hydrocarbons, e.g. those emitted from industrialised sites in the Northern Hemisphere. It should be noted however, that not all VOCs are anthropogenic. For example, isoprene enters the atmosphere at the highest rate and is the most abundant biogenic VOC, but is emitted from trees during photosynthesis rather than from any anthropogenic source. On a global scale, biogenic emissions of VOCs dominate, but anthropogenic emissions account for 10–20% of the total, and dominate in urban areas.

We conclude our discussion of tropospheric oxidation chemistry by emphasising that although the oxidation of methane has been used as a generic example of the oxidation cycle, there is a vast array of more complex VOCs present in the atmosphere, both as a result of direct emissions and formed as the products of oxidation chemistry. Consider for example peroxy radicals, RO_2, which may undergo various reactions, including decomposition in the case of larger radicals of this type. RO_2 reacts with NO and HO_2 (the dominant sink) and other RO_2 (a minor sink). They also react with NO_2, but in general the products decompose rapidly, the exception being peroxyacyl radicals, which produce PAN. $RO_2 + HO_2$ produces organic hydroperoxides, ROOH, while $RO_2 + NO_2$

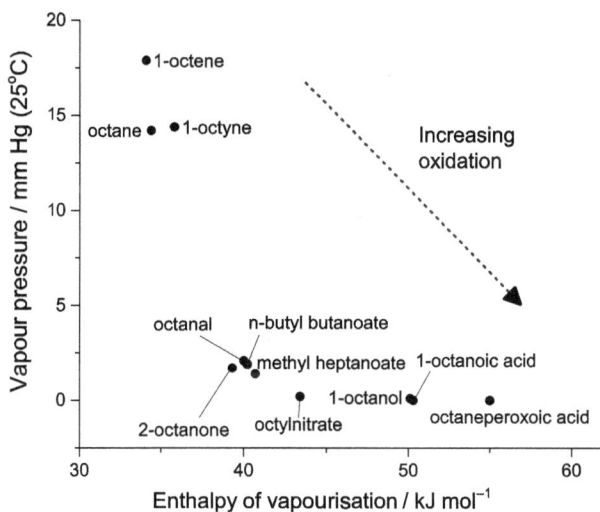

Figure 4.15   The vapour pressures of a series of functionalised C_8 organic molecules.

produces carbonyls (aldehydes and ketones) and also organic nitrates as a minor product. $RO_2 + RO_2$ reactions produce a range of oxygenated organic compounds, including carbonyls, carboxylic acids, alcohols, and esters. The different products can have very different vapour pressures, and many readily condense to form aerosol particles, with the partitioning of these VOCs between the gas and condensed phases being highly dependent upon the degree of oxygenation. For example, the effect of changing the oxidation of C_8 hydrocarbons on the vapour pressure is shown in Figure 4.15. Clearly, the detailed molecular structure has a profound effect on gas-particle partitioning. Aerosols and their importance in the atmosphere will be discussed in Chapter 5.

4.8 Questions

4.8.1 Essay-like Questions

Q4.1: Explain why the presence of methane in the troposphere is important for the oxidation of NO to NO_2 and thus for the formation of tropospheric ozone.

Q4.2: Draw up a chemical reaction scheme showing the elementary steps involved in the oxidation of methane in the troposphere. Identify all radicals and other intermediates that appear in your scheme.

Q4.3: (a) The daytime chemistry of the Earth's atmosphere differs markedly from that at night. Explain why this is so, giving, examples of some of the processes which are affected. (b) Ozone photolysis is an important process in the atmosphere, with an onset at $\lambda \approx 310$ nm. One of the products is $O(^1D)$; what is the other product, in what state is it principally formed and why?

Q4.4: Within tropospheric chemistry, what is meant by (i) the NO_x-limited regime, and (ii) the VOC-limited regime? Will reducing NO_x levels always lead to a reduction in ozone levels?

4.8.2 Problems

P4.1: The reactions

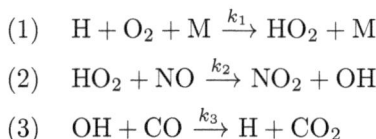

$$(1) \quad H + O_2 + M \xrightarrow{k_1} HO_2 + M$$
$$(2) \quad HO_2 + NO \xrightarrow{k_2} NO_2 + OH$$
$$(3) \quad OH + CO \xrightarrow{k_3} H + CO_2$$

are the most important in determining the ratio $[OH]/[HO_2]$ in the troposphere.

(a) What is meant by "M" in Equation (1), and what part does it play in the reaction?

(b) Calculate the steady-state concentration of HO_2 for an air parcel in which $[OH] = 1.6 \times 10^6$ molecules cm^{-3}, $[CO] = 3 \times 10^{12}$ molecules cm^{-3} and $[NO] = 1 \times 10^9$ molecules cm^{-3}, given that $k_2 = 8.1 \times 10^{-12}$ cm^3 molecule^{-1} s^{-1} and $k_3 = 2.4 \times 10^{-13}$ cm^3 molecule^{-1} s^{-1}. Assume that reaction (1) is sufficiently rapid in the regions of the atmosphere under discussion to convert H to HO_2 instantaneously.

(c) The rate constant for the reaction of OH with CH_4 is 4×10^{-15} cm^3 molecule^{-1} s^{-1} at the average temperature of the troposphere (277 K). Calculate the lifetime of methane in the atmosphere. Assume that [OH] is constant during the day, and averaged over a 24-hour period (day and night) is half that given in part (b).

P4.2: Measurements of methane in the atmosphere show that it is present at a volume mixing ratio of 1.8×10^{-6}. Assuming that the total mass of air in the troposphere is 4.5×10^{18} kg and that the average molar mass of air is $28.8 \, \mathrm{g \, mol^{-1}}$, estimate the source strength (annual release) of CH_4 in units of kg yr^{-1}. State one natural and one anthropogenic source of the hydrocarbon.

P4.3: (a) During the daytime, a typical rate of absorption of radiation with $\lambda < 310$ nm by ozone in the lower atmosphere is 2.1×10^8 photons cm^{-3} s^{-1} and the quantum yield for formation of $O(^1D)$ is 0.9. The $O(^1D)$ atoms are removed primarily by two processes:

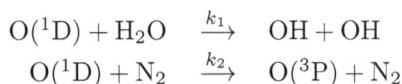

$$O(^1D) + H_2O \xrightarrow{k_1} OH + OH$$
$$O(^1D) + N_2 \xrightarrow{k_2} O(^3P) + N_2$$

Calculate the rate of formation of OH radicals given that $k_1 = 2.2 \times 10^{-10}$ cm^3 molecule^{-1} s^{-1} and $k_2 = 1.8 \times 10^{-11}$ cm^3 molecule^{-1} s^{-1} at the appropriate temperature and that the mole fractions of H_2O and N_2 are 0.02 and 0.78 respectively. If the hydroxyl radical is removed in reactions for which the rate is given by the expression 60[OH] molecule cm^{-3} s^{-1}, estimate the mean day-time concentration of OH under these conditions, given that its typical mole fraction is $\ll 6 \times 10^{-11}$.

(b) Isoprene is an important bio-organic hydrocarbon released by many types of vegetation. The rate constant for its reaction with the OH radical is 1.0×10^{-10} cm^3 molecule^{-1} s^{-1}. Calculate the lifetime of isoprene during the day from this removal process.

(c) At night, attack on isoprene is mainly due to the reaction with ozone and the nitrate radical, NO_3. The rate constants for the reactions with these two species are 1.3×10^{-17} and 6.8×10^{-15} cm^3 molecule^{-1} s^{-1} respectively, and the mole fractions of O_3 and NO_3 are 8×10^{-8} and 6×10^{-11}. If the supply of isoprene is suddenly cut off (for example, by a change in wind direction), calculate the fraction of isoprene removed near the ground by these two reactions during a 12-hour night. [The number density at standard temperature and pressure (STP) is 2.5×10^{19} cm^{-3}].

P4.4: (a) Briefly describe the principles behind CRDS.

(b) The ring-down time, τ, for a cavity of length L formed by mirrors of reflectivity R and enclosing a homogeneous gas sample at concentration $[A]$ is:

$$\tau = \frac{L}{c(1 - R + \sigma[A]L)}$$

where c is the speed of light and σ is the absorption cross-section of the sample. The visible absorption spectrum of the NO_3 radical has a maximum absorption cross-section of value 2.2×10^{-17} cm^2 molecule^{-1} at 662 nm. If a CRDS measurement conducted in a 1 m long cavity displays a background ring-down time of $\tau_o = 180\,\mu s$ and a minimum detectable change of $\Delta\tau = 0.1\,\mu s$, calculate:

(i) the reflectivity, R, of the cavity mirrors;

(ii) the detection limit for NO_3 in molecules cm^{-3}.

(c) Discuss the role of NO_3 in the night-time chemistry of the atmosphere.

P4.5: In many regions of the troposphere, the ozone mixing ratio is controlled by the following three chemical reactions:

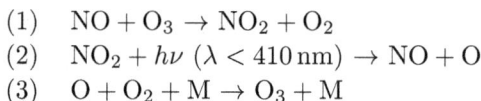

(1) $NO + O_3 \rightarrow NO_2 + O_2$
(2) $NO_2 + h\nu\ (\lambda < 410\,nm) \rightarrow NO + O$
(3) $O + O_2 + M \rightarrow O_3 + M$

(a) Given that reaction (1) has a bimolecular rate constant k_1 and reaction (2) has a rate $J_2[NO_2]$, and assuming that NO_2 is in steady state, obtain an expression for the concentration of ozone, $[O_3]$, in terms of $[NO]$ and $[NO_2]$.

(b) Hence calculate the steady-state mixing ratio of ozone (in parts per billion by volume, ppbv) at midday, given that the total pressure is 101.3 kPa, the temperature is 298 K, $k_1 = 1.8 \times 10^{-14}$ cm^3 molecule^{-1}s^{-1}, and $J_2 = 10^{-2}$ s^{-1}. Typical tropospheric mixing ratios for NO and NO_2 are 5 and 10 pptv (parts per trillion by volume), respectively.

(c) Explain why the concentration of ozone increases when hydrocarbons are present in the polluted daytime troposphere.

P4.6: In this problem we consider the rate of ozone production in the presence of RO_2 and NO_x from the following set of reactions:

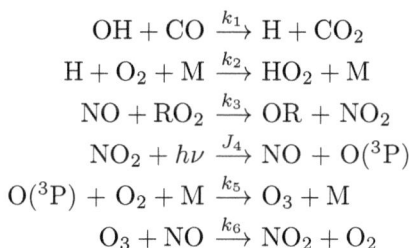

$$OH + CO \xrightarrow{k_1} H + CO_2$$
$$H + O_2 + M \xrightarrow{k_2} HO_2 + M$$
$$NO + RO_2 \xrightarrow{k_3} OR + NO_2$$
$$NO_2 + h\nu \xrightarrow{J_4} NO + O(^3P)$$
$$O(^3P) + O_2 + M \xrightarrow{k_5} O_3 + M$$
$$O_3 + NO \xrightarrow{k_6} NO_2 + O_2$$

(a) Derive an expression for $d[O_3]/dt$.
(b) Assume that NO and O are in steady state and then substitute into your answer for (a) to show that

$$\frac{d[O_3]}{dt} = J_4[NO_2]\left(1 - \frac{k_6[O_3]}{k_6[O_3] + k_3[RO_2]}\right)$$

where $RO_2 = HO_2 + CH_3O_2$.
(c) Identify the value of $[RO_2]$ for which $d[O_3]/dt = 0$.
(d) As $[RO_2] > 0$ then O_3 is produced. Furthermore, increasing NO_x levels leads to increasing O_3 production. An important question to consider is for what value of $[NO]$ does the atmospheric chemistry change from being net ozone destroying to net ozone producing?

For a first approximation, we consider HO_2 to be the only peroxy radical present. The transition occurs when there is sufficient NO for the reaction $NO + HO_2$ to be competitive with the $O_3 + HO_2$ reaction:

$$O_3 + HO_2 \xrightarrow{k_7} OH + 2O_2$$
$$NO + HO_2 \xrightarrow{k_3} OH + NO_2$$

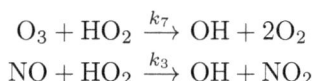

Given that a typical value of $[O_3]$ is 40 ppbv, and k_3 and k_7 take values 7.7×10^{-12} and $1.9 \times 10^{-15} \, cm^3 \, molecule^{-1} \, s^{-1}$ respectively, estimate the minimum value of $[NO]$ that ensures net ozone production. Comment on your result.

Chapter 5

Aerosols and Heterogeneous Reactions

Introduction

Atmospheric aerosols consist of an ensemble of microscopic particles suspended in air, and may be solid, liquid or multiphase in nature. Aerosols can provide a substrate for chemical reactions, and therefore play an important role in the overall chemistry of the atmosphere, with some reaction cycles involving both gas-phase and surface-catalysed reactions (Chapter 3). In addition, many of the gas-phase oxidation products formed in the troposphere (Chapter 4) are removed from the atmosphere by uptake into aerosols and clouds, and are hence physically removed by dry or wet deposition. The existence of atmospheric aerosols can impact upon climate, both directly through absorption and scattering of solar and terrestrial radiation (Chapter 2), and indirectly by their ability to nucleate cloud droplets or ice particles, or to change cloud properties depending upon chemical composition. Furthermore, aerosols have a deleterious effect on air quality, not only reducing visibility but also exhibiting toxicity. As a result, the emission of particulate matter (PM) with radius less than 2.5 μm is legally regulated. These various diverse aspects of aerosol chemistry are highlighted in the present chapter.

5.1 The Aerosol Size Distribution

Aerosols have a wide size distribution, ranging from *ca.* 1 nm to 10 μm. A generic aerosol size distribution is shown in Figure 5.1. By convention, the distribution is divided into three regions, corresponding to: (i) *nucleation mode*, where the particle diameter, D, is less than 100 nm, (ii) *accumulation*

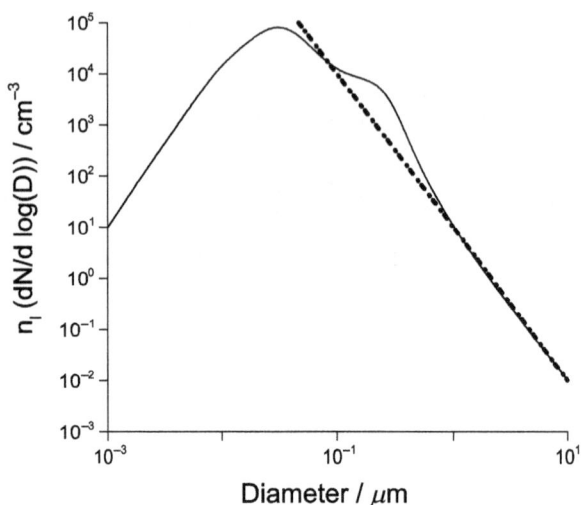

Figure 5.1   A generalised aerosol size distribution.

mode, where $100\,\mathrm{nm} < D < 2\mu\mathrm{m}$, and (iii) *coarse mode*, where $D > 2\,\mu\mathrm{m}$. The distribution of aerosol sizes results from the physical and chemical processes by which they are generated and lost. For example, the smallest particles are produced by gas-to-particle conversion processes, including condensation of hot vapours and chemical conversion of gases to low volatility vapours. In contrast, the largest particles are generated by the mechanical action of wind on the Earths' surface. The largest size of particle formed is determined by the *settling velocity* (or terminal velocity) of the aerosol particles: those with diameters much greater than 10 μm are likely to settle out in about a day or less and are rarely found in the atmosphere, except locally for short periods of time, for example in thunderstorm outflows. The size distribution also varies with locale, as it depends on both the source and loss mechanisms operating in a given region.

   The location of the maximum in the size distribution is related to the age of the aerosol: freshly formed aerosols, resulting from gas-to-particle conversion, are dominated by the smallest particles. After some time, *coagulation* of smaller particles into larger ones creates a maximum at intermediate sizes. Further coagulation tends to shift the maximum to larger diameters. At the same time, removal of the larger particles by sedimentation, rainout (in which aerosol particles act as cloud condensation nuclei, CCN), and washout (removal by raindrop and snowflake scavenging), lowers the concentration of large sizes. Loss by deposition depends on particle size: very large particles are rapidly lost due to sedimentation and impaction, while very small particles are rapidly lost by diffusion to other surfaces. When all of these factors are taken into account, it is found that the particles with the longest lifetime are those in the middle of the accumulation mode region, i.e. with diameters *ca.* 0.1–2 μm.

The size distribution can be defined by the *number density function*, $n(D)$, such that $n(D)dD$ is the number of aerosol particles per unit volume in the size range D to $D + dD$. The number of aerosol particles per unit volume of size $\geq D$ is then given by the *number distribution function, $N(D)$*.

$$N(D) = \int_D^\infty n(D)dD \qquad (5.1)$$

i.e. $n(D) = -dN/dD$. Because atmospheric aerosol sizes range over several orders of magnitude, it is convenient to plot the size distribution as a function of the logarithm of particle diameter. Hence we can define the *logarithmic number density function $n_l(D)$* as

$$n_l(D) = -dN/d(\log D) \qquad (5.2)$$

As noted previously, the aerosol number density depends, not surprisingly, on the spatial location where measurements are made, and the total number density generally decreases as one moves from continental to maritime air masses. As shown in Figure 5.2, the total number density of aerosols also generally decreases as one moves away from the Earth's surface. Background concentrations of small accumulation mode particles range from 10 to 100 cm^{-3} in the free troposphere. Typical concentrations of aerosol particles over the ocean are around 300 cm^{-3}, but vary by as much as a factor of three. Over the continent, concentrations are much higher, with concentrations of around 10^4 cm^{-3} near urban areas, approaching 10 times this in very polluted regions. These number densities are summarised in Table 5.1. Despite the temporal and spatial dependence of the aerosol size distributions, average size distributions tend to exhibit one robust feature, namely a slope of –3 for the logarithmic density function to the right of its maximum. Thus, for $D > 0.2$ μm the logarithmic density function behaves approximately as $n_l(D) = k/D^3$; this is known as the *Junge distribution*.

The primary production routes for aerosol particles are essentially limited to processes occurring at the Earth's surface, and hence the bulk of the aerosol mass is concentrated in the planetary boundary layer.

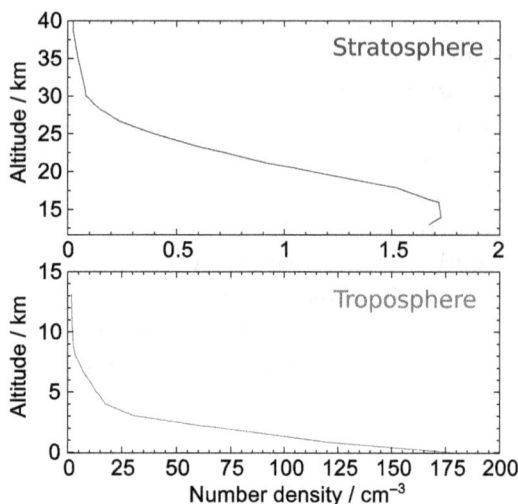

Figure 5.2   An example of the variation in aerosol number density with altitude. The profile refers to a background load of stratospheric aerosol.

Source: Adapted from B. Ovigneur *et al.*, *Atmos. Meas. Tech.* **4** 2359 (2011).

Table 5.1   Typical range of aerosol number densities in various geographical locations.

Location	Aerosol number density/cm^{-3}
Urban	5×10^3–5×10^5
Rural	10^3–5×10^4
Ocean	10^2–10^3
Mountain ($> 2\,$km)	10^2–10^3

This lowest kilometre or so of the atmosphere contains as much as 80% of the aerosol mass. Assuming a Junge distribution, about one third of the aerosol mass corresponds to the coarse mode size region, and two thirds to the fine mode (the sum of nucleation and accumulation modes). The coarse mode, with its greater settling velocity, is more concentrated close to its sources at the surface, whereas fine mode aerosol concentrations are also more pronounced near the surface, but less markedly so. At altitudes of between 5 and 10 km, the distribution of nucleation mode particles is rather uniform. Accumulation mode particles follow the profile of the smaller nucleation particles through the troposphere, but have a concentration maximum in the lower stratosphere in a layer between 15 and 25 km.

Within this *Junge layer*, the accumulation mode particles show a maximum concentration of *ca.* $10 \, \text{cm}^{-3}$ during periods of volcanic inactivity.

Globally, it is estimated that 75% of the aerosol mass is in the form of *primary* aerosols, consisting of sea spray (40%), mineral, volcanic and anthropogenic dust (20%), and particles produced in combustion (15%). The remaining 25% of the aerosol mass consists of *secondary* particles arising from gas-to-particle conversion. The principal constituents of atmospheric aerosols are quartz, organic materials and the ions Na^+, Cl^-, NH_4^+, SO_4^-, NO_3^-; aerosol composition is discussed further in the next section.

In some cases, it is not the number distribution that it is important, but rather the surface area or the volume distribution functions, respectively. These are shown in Figure 5.3. For example, the surface area distribution is of importance to atmospheric chemistry, since the rate of chemical reactions taking place on the surface of aerosols will depend on their total surface area. If the mass of a pollutant that is being transported is of interest, then the mean diameter of the particles with a given volume (assuming that density is independent of particle size) needs to be determined. When considering the implications of various particle sizes for human health, both the mass and the number of the particles are important, since only certain sizes of particle can enter the lungs. The number, surface area and volume distributions are related as follows:

$$N = \int_{-\infty}^{\infty} n_l(\log D) d \log D \tag{5.3}$$

$$S = \pi \int_{-\infty}^{\infty} n_l(\log D) D^2 d \log D \tag{5.4}$$

$$V = \frac{\pi}{6} \int_{-\infty}^{\infty} n_l(\log D) D^3 d \log D \tag{5.5}$$

5.2 Aerosol Composition

It should be clear from the discussion above that aerosols come in all shapes, sizes and compositions. In this section, the main primary aerosol types are summarised. As mentioned above, aerosol production mechanisms are classified as either *primary* or *secondary*. Primary production refers to mechanisms that mobilise PM directly, for instance from the bulk. Examples include erosion of the land surface, the decomposition of biomass,

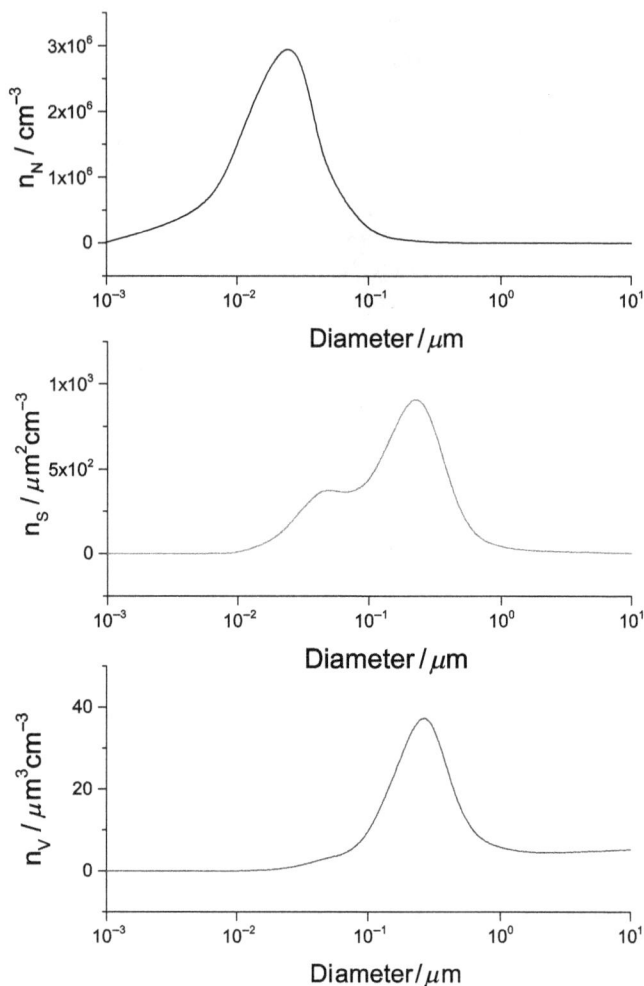

Figure 5.3   Exemplar aerosol number, surface, and volume distributions.

and sea spray. Secondary production refers to the formation of particles from the gas phase, through condensation of vapour. Primary production of aerosols can produce particles of all size ranges, with dust and sea-salt particles classified as coarse mode aerosols. The primary aerosol types are (i) sea-salt particles, (ii) sulphates, (iii) fine dust, (iv) carbon particles (soot), and (v) other organics. Each type will be discussed further below. Figure 5.4 shows a summary of the chemical composition of aerosols as a function of geographic location.

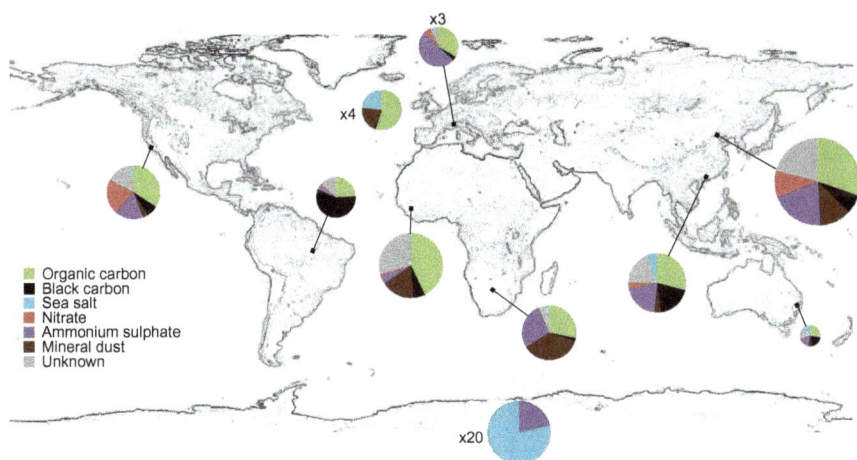

Figure 5.4 Aerosol composition at different locations. The area of the pie charts represents relative concentrations of fine aerosol, with scaling factors indicated.

Source: Data from: Chan 1997 (Brisbane, Long Beach); Fuzzi 2007 (Brazil); Jourdain 2002 (Antarctica); Qin 1997 (Hong Kong); Cozic 2008 (Jungfraujoch); He 2001 (Beijing); Puxbaum 2000 (South Africa); Weinstein 2010 (Guinea); Cavalli 2004 (North Atlantic).

Sea-salt aerosols are formed by the bursting of spray droplets formed by wave action at the ocean surface. The roughening of the sea is important here because breaking waves force air into the water, which then nucleates bubbles that move to the surface and break. NaCl is very hygroscopic, and NaCl aerosols act effectively as *cloud condensation nuclei*, a topic which we will return to later. Sea salt has a wide size distribution $(0.05–10\,\mu m$ diameter) and as such has a correspondingly diverse atmospheric lifetime.

Emission of SO_2 from the burning of sulphur-containing fossil fuels accounts for *ca.* 70% of all sulphur emission into the atmosphere. SO_2 is readily oxidised to SO_3 then H_2SO_4, which condenses with water to form $H_2SO_{4(aq)}$ droplets. H_2SO_4 is then neutralised by NH_3 in the reaction

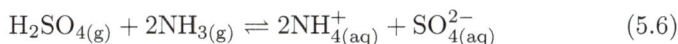

$$H_2SO_{4(g)} + 2NH_{3(g)} \rightleftharpoons 2NH_{4(aq)}^+ + SO_{4(aq)}^{2-} \tag{5.6}$$

to yield $(NH_4)_2SO_{4\,(aq)}$, which is the predominant form of sulphate aerosols. The remainder are formed from volcanic and oceanic emissions of reduced sulphur gas (H_2S and dimethyl sulphide), which are relatively quickly oxidised to H_2SO_4. The residence time for tropospheric aerosols is around 1–2 weeks, but some of the sulphate aerosols formed at higher altitude have a longer residence time. Sulphate aerosols tend to be very hygroscopic, and thus are very effective CCN.

In an analogous manner, nitrate also forms aerosols efficiently, via reaction with ammonia at neutral pH.

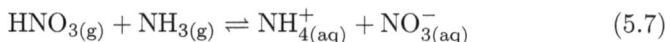

$$HNO_{3(g)} + NH_{3(g)} \rightleftharpoons NH_{4(aq)}^+ + NO_{3(aq)}^- \tag{5.7}$$

The resulting ammonium nitrate aerosol is ubiquitous and is a very efficient absorber of long λ radiation.

The term *mineral dust* generally refers to silicate particles mobilised from the land surface. Sources of mineral dust are deserts, semi-arid regions, and agricultural regions during soil disturbance. As discussed previously, the residence time depends on the size of the particles; the largest particles sediment quickly while sub-micron particles can be transported over long distances. Dust particles tend to be relatively large, and because they are often insoluble, they may function as effective ice nuclei.

Black carbon is largely elemental carbon in the form of soot, and is produced from incomplete natural or anthropogenic biomass burning, as well as industrial combustion processes. Although the mass of black carbon in the atmosphere is one to two orders of magnitude less than that of sea salt, it is extremely important because it absorbs visible radiation, increasing the radiative impact of atmospheric aerosols — see Section 5.4.

Organic aerosol is a broad term covering the wide variety of organic matter present in the atmosphere and is distinguished from black carbon on the basis of its light absorption and scattering properties. Organic aerosol particles, or organic coatings of other aerosols, is responsible for more than 50% of the total aerosol mass over continental areas, and appear to be oxygenated and polar (having low vapour pressure), being particularly derived from carboxylic acids. Much of the organic aerosol is the result of condensation of the corresponding gas-phase species originating from the terrestrial biosphere. However, bulk production of organic PM from decaying plants and by the emission of pollen is also important. Organic aerosol provides a direct link between the biosphere and atmospheric processes.

While the discussion above focuses primarily on tropospheric aerosols, it should be noted that the Junge layer comprises a layer of aerosol particles residing in the lower stratosphere from the tropopause up to an altitude of *ca.* 30 km. The particles have a mean diameter of *ca.* 100 nm and are produced by the condensation of sulphuric acid with a small amount of water. The sulphuric acid itself is produced from the oxidation of SO_2, OCS (produced in the ocean), and transport of aerosols from the troposphere. In the absence of volcanic emissions this aerosol has negligible optical depth, though can greatly influence ozone chemistry. Volcanic eruption can however increase the stratospheric aerosol optical depths by orders of magnitude. Volcanic eruptions inject large amounts of SO_2 directly into the stratosphere. Most of the volcanic ash is sufficiently large that it falls quickly. However, SO_2 persists, and is oxidised over a period of 3–6 months by the OH radical in the following reaction sequence:

$$OH + SO_2 + M \rightarrow HSO_3 + M$$

$$HSO_3 + O_2 \rightarrow SO_3 + HO_2$$

$$SO_3 + H_2O \rightarrow H_2SO_4$$

The resultant H_2SO_4, then accumulates on the aerosol. These aerosols have a residence time of 1–2 years due to their low rates of sedimentation and mixing back into the troposphere where they are scavenged by surfaces and rain. Sulphate has a number of strong absorption features between 3 and $20\,\mu m$, and in addition to reflecting solar radiation, these aerosols absorb terrestrial radiation efficiently, heating the stratosphere and producing a small greenhouse effect.

5.3 Aerosols and the Formation of Clouds

As a prelude to a discussion of the role of clouds and aerosols in radiative forcing within the atmosphere, we now consider the mechanism by which aerosols facilitate cloud formation.

5.3.1 The Vapour Pressure of Water Droplets: Kelvin's Equation

In the absence of particles, condensation from the gas to the liquid phase proceeds upon randomly formed (micro)droplets, a process known as *homogeneous nucleation*. While condensation decreases the Gibbs free energy, G, of the gas-droplet system through the release of heat, there is an accompanying increase in G associated with the increase of the surface energy of the water droplet. Kelvin showed that at equilibrium, the saturated vapour pressure, p, over a spherical droplet of radius r is

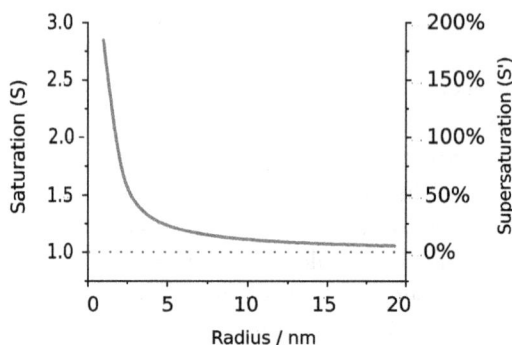

Figure 5.5   Vapour pressure over curved (solid line) and flat (dotted line) surfaces for pure H_2O at 298 K.

given by

$$p(r) = p_\infty \exp\left(\frac{2\gamma}{r} \frac{M}{\rho RT}\right) \qquad (5.8)$$

where p_∞ is the saturated vapour pressure over a plane surface, γ is the surface tension of the liquid from which the droplet is formed, and M and ρ are the molar mass and density of the liquid respectively. In the case of water, $\gamma = 72 \times 10^{-3}\,\mathrm{J\,m^{-2}}$ at 298 K. The Kelvin equation shows that the vapour pressure above the droplet is always larger than for a plane surface. This behaviour is shown in Figure 5.5, in which the *saturation*, S, defined as $p(r)/p_\infty$, is plotted as a function of droplet radius. Also plotted is the *supersaturation*, S', defined as $(S - 1) \times 100\%$.

The Kelvin equation describes an *unstable equilibrium*, in the sense that a droplet of pure water of fixed radius can only be stable under the condition that the water vapour pressure of the surrounding air, p_{env}, is equal to $p(r)$. A small droplet of pure water will therefore evaporate unless the surrounding air is highly supersaturated. Furthermore, for surroundings with a given saturation, S^*, the droplet must be formed with some *critical radius*, $r^* = 2M\gamma/(\rho RT \ln S^*)$. For example, if the surroundings have a saturation of 1.1, i.e. a supersaturation of 10%, then the critical radius is 10 nm. Formation of a droplet of this diameter would require the simultaneous collision of around 10^5 water molecules. The probability of such homogeneous nucleation is vanishingly small, and it can be concluded that an alternative mechanism for the formation of water droplets must operate. In fact, it is the presence of wettable or water-soluble aerosol particles in the atmosphere that facilitates the formation of large droplets.

As we will see in the next sections, the presence of only a very small amount of aerosol (10^{-15} g or less) is sufficient to reduce the saturated vapour pressure above the droplet's surface so that droplet formation/growth occurs; if aerosols are wettable, water vapour can condense on them and droplet growth can occur at radii well above r^*.

5.3.2 Water Vapour and Aerosols: Hygroscopic Growth

Water vapour is ubiquitous in the lower troposphere, with the result that layers of water are present even on 'dry' particles. The amount of water on the surface of an aerosol particle depends upon both the *hygroscopicity* (ability to take up water) of the particle and on the relative humidity (RH) of the air. Clouds form through condensation of water vapour onto hygroscopic particles known as *cloud condensation nuclei* (CCN). The effectiveness of an aerosol particle as a CCN depends upon its hygroscopicity, size, composition, and phase. The change of a solid soluble aerosol to a liquid aerosol is called *deliquescence* while the reverse process is known as *efflorescence*. Cloud droplets form on aqueous aerosol particles, meaning that aerosol particles have taken up water prior to being activated into droplets. We will therefore first discuss the different phases of soluble aerosol particles before considering cloud droplet formation.

The growth of an aerosol particle is quantified by the *growth factor*, $g(\text{RH})$, defined as the ratio of the particle diameter to its diameter at 0% RH. The growth factor depends on the aerosol particle size through the Kelvin equation as shown in Figure 5.6(a). Here the growth of two H_2SO_4 particles (with different initial radii, D_P) in response to increasing RH are compared with theory. Some aerosols, such as H_2SO_4, remain liquid at all RHs and only change size according to RH; for H_2SO_4 this is a consequence of the relatively small saturation vapour pressure. Other aerosols, such as NaCl and $(NH_4)_2SO_4$, change their phase with changing RH. Importantly, the RH at which the phase change occurs depends on the *direction* of the change in RH. For example, consider the case of a pure NaCl particle[1] (dashed lines in Figure 5.6(b)) whose (dry) mass at 0% RH is denoted w_0. As the RH increases, the particle mass remains constant up to 75% RH.

---

[1] Interestingly, while sea-salt particles both grow and evaporate in a similar manner to the pure NaCl particles/droplets, there are subtle differences which are due to fact that sea salt is a complex mixture of inorganic salts and may also contain organic compounds.

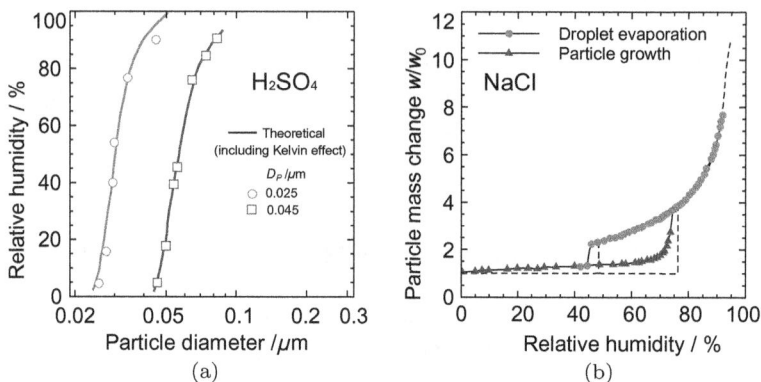

Figure 5.6   Water uptake by (a) H_2SO_4, and (b) sea-salt particles (NaCl). In (b) the hydration behaviour of a pure NaCl particle is illustrated as dashed line.

Source: Adapted from (a) B.Y.H. Liu *et al.*, *Atmos. Environ.* **12** 99 (1978), and (b) I.N. Tang *et al.*, *J. Geophys. Res.* **102** 23269 (1997).

At this point, the mass, w, (and the diameter) suddenly increases as the particle takes up water and changes phase from solid to liquid, i.e. the salt dissolves in water. This *deliquescence RH* (DRH) occurs when a sufficient number of water molecules are available to dissolve the salt and form a saturated solution. For a pure substance, the DRH is determined by the solubility of the salt in water. If the RH is now decreased, NaCl remains in the liquid phase until *ca.* 45% RH, where it solidifies; this defines the *crystallisation RH* (CRH). g(RH) for NaCl clearly exhibits *hysteresis*: the growth factor is not unambiguously determined by RH, but depends also upon the history of the particle. Between 45 and 75% RH the liquid phase of NaCl is *metastable* with respect to the solid phase.

The hysteresis may be explained further by considering the Gibbs free energy as a function of RH for a generic NaCl aerosol with a DRH of 75%. For a solid, G $(= U + PV - TS)$ is independent of RH and is therefore a constant (internal energy, pressure, volume, temperature, and entropy are all independent of RH). In contrast, G for an aqueous droplet decreases with increasing RH. This is because the entropy, S, of the droplet increases as its volume increases. The thermodynamically-favoured phase is that with the lowest Gibbs free energy, with the consequence that when the RH is higher than the DRH, the aqueous phase is the more stable state, while below the DRH the solid is more stable. Nevertheless, the NaCl aerosol remains liquid down to the crystallisation RH when RH decreases. This is because although the phase change is thermodynamically favoured, whether the phase transition occurs on an observable timescale depends not only upon the free energies of the initial and final state, but also on G for all the intermediate states that the system must go through. In the case of a salt solution droplet at a RH below the DRH, transition to the solid state involves the formation and growth of a solid *embryo* within the aqueous solution. The formation of a solid embryo requires the creation of a surface, which requires energy. This energy barrier needs to be overcome before the aqueous aerosol can crystallise, i.e. the phase change is kinetically inhibited. A phase change in which a thermodynamically stable phase grows within the metastable phase is called *nucleation*. In Figure 5.6(b), the intermediate supersaturated state persists down to 45% RH, where the aqueous solution is sufficiently supersaturated for crystallisation to occur. In essence, efflorescence requires a nucleation process whereas deliquescence does not.

5.3.3 The Role of Solute in Cloud Formation: Raoult's Law

Raoult's law states that the vapour pressure of a solvent above an ideal solution is proportional to its mole fraction within the solution. Hence, the presence of a solute within a droplet lowers the water vapour pressure above the droplet. The ratio of water vapour pressures for a droplet containing dissolved solute compared to that for a pure droplet is then $(n_w/(n_w+n_s))$ where n_w and n_s are the number of water and solute molecules respectively. The ratio $(n_w/(n_w+n_s))$ is commonly known as the *water activity*, a_w. For a dilute solution, $n_w \gg n_s$, and so

$$a_w = (1 + (n_s/n_w))^{-1} \approx 1 - (n_s/n_w) \qquad (5.9)$$

Since the number of water molecules in a spherical droplet is proportional to r^3, the water activity of the aqueous droplet of radius, r, can be expressed as

$$a_w = 1 - \left(\frac{b}{r^3}\right) \qquad (5.10)$$

where the constant b is proportional to the amount of solute in the droplet, and also accounts for the degree of ionic dissociation of the solute in solution.[2]

---

[2]It can be shown that $b = \frac{3iM_w m_s}{4\pi\rho_l M_s}$ where m_s is the mass of the solute, M_s and M_w are the molecular masses of the solute and water respectively, ρ_l is the density of the droplet, and i is the *Van't Hoff factor* which accounts for ionic dissociation of the solute.

This result shows that for a fixed mass of solute, the smaller the droplet, the larger the effect of the solute molecules in reducing the ambient vapour pressure required for the droplet to be in equilibrium with its surroundings. On the other hand, the curvature of the droplet, via the Kelvin effect, progressively increases the equilibrium vapour pressure for decreasing droplet sizes. The balance between these effects is discussed in the next section.

5.3.4 Köhler Curves

The saturation ratio, S, over a droplet is therefore the product of two terms; a 'curvature' term which increases the saturation as droplet size decreases, and a 'solute' term which tends to decrease it:

$$S(r) = \exp\left(\frac{a}{r}\right)\left(1 + \frac{b}{r^3}\right)^{-1} \tag{5.11}$$

As $a \ll r$ and $b \ll r^3$, expanding each term in $S(r)$ and neglecting higher order terms yields $S(r) \approx (1 + (a/r) - (b/r^3))$.[3] A plot of $S(r)$ is known as a Köhler curve, and is shown in Figure 5.7, along with the separate curvature and solute effects. The plot displays a maximum, S_c, at the critical radius, $r_c = (3b/a)^{1/2}$. Typical values for the constants a and b are $a = 6$ nm and $b = 2 \times 10^{-5}$ (μm^3), and hence r_c is typically around 100 nm. If a particle

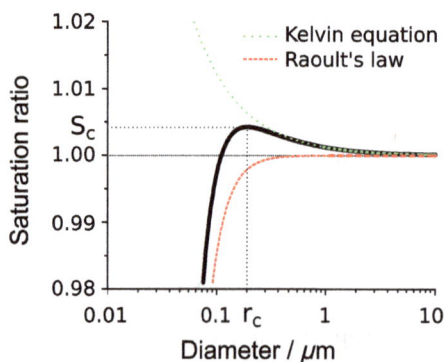

Figure 5.7   An exemplar Köhler curve. Assumed $M = 58$ g/mol, $m = 10^{-16}$ g, $i = 2$, $T = 273$ K. The dashes indicate the contribution from Raoult's Law and the dots that from the Kelvin equation.

---

[3]For $x \ll 1$, we can make the approximations $e^x \approx (1 + x)$ and $(1 + x)^{-1} \approx (1 - x)$.

of this radius either grows or shrinks then the vapour pressure above its surface will be reduced. In practice, a smaller droplet will tend to shrink and a larger one to grow.

If the saturation ratio above the droplet, S, is below the critical ratio S_c, the droplet can only grow to the radius at which the Köhler curve takes the value S. Under these conditions the droplet will be stable. Such droplets give rise to *haze* when they form in the atmosphere. The stable equilibrium leading to the production of such droplets can be appreciated as follows. We assume that the droplet is in equilibrium with its surroundings, i.e. the ambient water vapour pressure, p_{env}, is the same as that above the droplet, $p(r)$. If the size of the droplet increases slightly then its equilibrium vapour pressure (described by the Köhler curve) will be higher than the (constant) pressure of the surroundings. The surroundings are then subsaturated with respect to the droplet, with the result that molecules will evaporate from the droplet until the droplet is again at equilibrium with its surroundings. A similar argument may be made for the case of the droplet size is decreased — the surroundings will then be supersaturated with respect to the droplet and water molecules will condense onto the droplet until equilibrium is re-established. The equilibrium droplet size is therefore that for which the saturation (predicted by the Köhler curve) is equal to the ambient saturation.

In contrast, if the ambient saturation ratio is greater than S_c, the droplet will continue to grow spontaneously by condensation. A droplet whose radius is greater than r_c (typically 0.1–1 μm) is called *activated*. As the droplet grows, the quantity $(S - S_c)$ increases and the surroundings become increasingly supersaturated for the growing cloud droplet, i.e. the water vapour concentration gradient between the droplet surface and the surroundings increases. In the context of cloud formation, the number of activated CCN is always small, *ca.* 200 cm^{-3}, compared with the total number of nuclei, which can be an order of magnitude larger. The most common natural source of nuclei is sea spray. It should also be noted that r_c is completely distinct from the r^* derived from the Kelvin equation. Furthermore, the Köhler equation pertains to the growth of solution droplets that originate from deliquescence of salt aerosol particles, i.e. the deliquescence is not described by the Köhler equation, but is assumed to have already taken place.

Figure 5.8 show a selection of Köhler curves for different masses of the solutes NaCl and Na$_2$SO$_4$. It is clear that the higher the ambient saturation ratio, S, the larger the number of droplets that can be activated

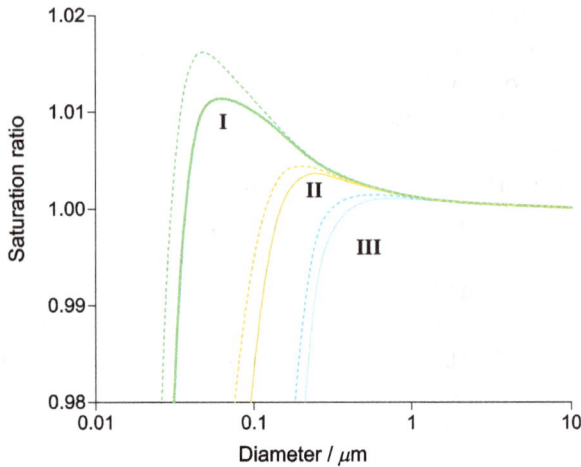

Figure 5.8   A series of Köhler curves. The different colours represent different solute masses: 10^{-17} g (I); 10^{-16} g (II); 10^{-15} g (III). Solid lines are calculated for NaCl and dashed for Na_2SO_4 (i.e. $M = 58$ u, $i = 2$; $M = 142$ u, $i = 3$). All assumed at $T = 273$ K.

and the smaller they can be. Larger aerosols particles can more readily act as CCN because they require small supersaturations to be activated, but there are fewer of them (see Section 5.1). As will be shown in the next section, the rate of droplet growth by diffusion of water molecules depends inversely on the radius of the droplet, and so large droplets may not reach their critical size by diffusional growth. In contrast, activation of nucleation mode aerosols requires very high supersaturations that do not exist in the atmosphere, and hence it is mainly accumulation mode aerosols that act as CCN and become activated to form cloud droplets. It is also clear from Figure 5.8 that all of the curves approach the Kelvin curve as the size of the droplet increases; this is because the droplets are becoming increasingly dilute solutions. Furthermore, at 100% RH, no aerosol can be activated to become a cloud droplet, as activation always requires supersaturation.

5.3.5   Rates of Droplet Growth

As long as the water vapour pressure in the surroundings exceeds the vapour pressure close to the droplet, the droplet can grow through condensation. The rate of growth is determined by the balance between condensation of vapour and the temperature rise due to latent heat release, which increases the evaporation rate. The growth therefore depends on the efficiency of heat transport from the droplet and the gradient in water vapour pressure.

For a droplet of mass m and radius r, the growth rate is given by

$$\frac{dm}{dt} = 4\pi r^2 D \left(\frac{d\rho}{dR}\right)_{R=r} \tag{5.12}$$

where D is the diffusion coefficient of water vapour in air and ρ is the vapour density, which is a function of radial distance R from the droplet. The boundary conditions are: (i) as $R \to \infty$ then $\rho \to \rho_\infty$, where ρ_∞ is the density of water vapour far from the droplet; and (ii) as $R \to r$ then $\rho \to \rho_r$, where ρ_r is the density at the droplet surface. The dependence of the vapour density on R is then

$$\rho(R) = \left[\rho_\infty - \left(\frac{r}{R}\right)(\rho_\infty - \rho_r)\right] \tag{5.13}$$

and hence

$$\left(\frac{d\rho}{dR}\right)_{R=r} = \left(\frac{1}{r}\right)(\rho_\infty - \rho_r) \tag{5.14}$$

Substituting into our earlier expression for the growth rate, we obtain

$$\frac{dm}{dt} = 4\pi r D(\rho_\infty - \rho_r) \tag{5.15}$$

The droplet therefore grows if $\rho_\infty > \rho_r$ and evaporates if $\rho_\infty < \rho_r$. While ρ_∞ is determined by ambient conditions, ρ_r depends on the droplet size, composition and temperature, T_r, which is not usually the same as the ambient temperature, T_∞ — see below. For a spherical droplet we also have

$$\frac{dm}{dt} = 4\pi r^2 \left(\frac{dr}{dt}\right)\rho_l \tag{5.16}$$

where ρ_l is the density of liquid water. Equating these last two expressions yields

$$\frac{dr}{dt} = \left(\frac{D}{r}\right)\left(\frac{\rho_\infty - \rho_r}{\rho_l}\right) \tag{5.17}$$

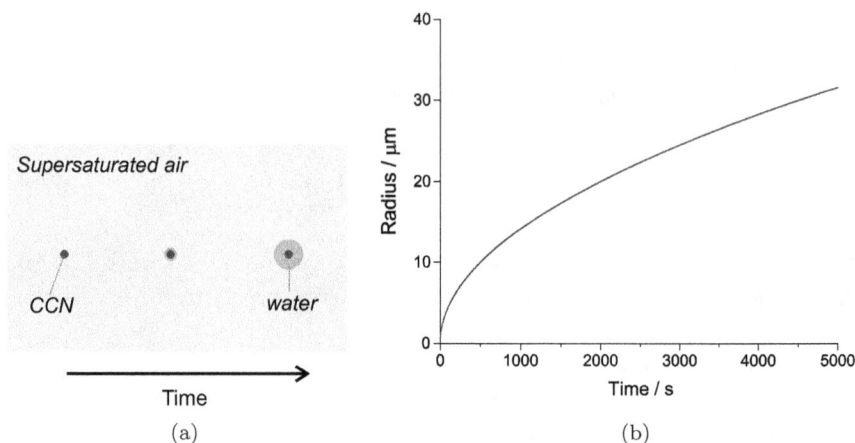

Figure 5.9   (a) Schematic for growth by condensation. (b) Droplet radius as a function of time under conditions of growth by condensation with $F = 10^{-2}\,\text{s}\,\mu\text{m}^{-2}$ and $S = 1.001$.

for the rate of growth of the droplet due to diffusion of vapour if its temperature, and hence ρ_r, remains constant. The droplet growth is therefore inversely proportional to the radius, as depicted in Figure 5.9(b).

The above expression assumes that the droplet remains at a constant temperature as it changes in size. However, in practice the droplet will gain heat Q due to the release of latent heat L at a rate given by

$$\frac{\mathrm{d}Q}{\mathrm{d}t} = L\frac{\mathrm{d}m}{\mathrm{d}t} = 4\pi r L D(\rho_\infty - \rho_r) \tag{5.18}$$

and will also lose heat due to the thermal conductivity of air, λ:

$$\frac{\mathrm{d}Q}{\mathrm{d}t} = 4\pi r^2 \lambda \left(\frac{\mathrm{d}T}{\mathrm{d}R}\right) = 4\pi r \lambda (T_\infty - T_r) \tag{5.19}$$

At equilibrium, these two terms balance, and the droplet is found to be warmer than its surroundings by an amount

$$(T_\infty - T_r) = L D \left(\frac{\rho_\infty - \rho_r}{\lambda}\right) \tag{5.20}$$

T_r and ρ_r are unknowns, but are related by the Köhler equation. The previous equation and the Köhler equation can then be solved numerically to obtain the growth rate under various conditions. For sufficiently large droplets, the following analytical solution for the growth is appropriate:

$$\frac{dr}{dt} = \left(\frac{1}{r}\right)\left[\frac{S-1}{F}\right] \tag{5.21}$$

where F is a (strongly) temperature- and pressure-dependent term accounting for the combination of latent heat release due to condensation and heat transport and the diffusion of water vapour onto the droplet; in general F decreases in size as the temperature is increased, meaning that the droplet will grow more rapidly at higher temperatures. As an example, taking $F = 10^{-2}\,\mathrm{s}\,\mu\mathrm{m}^{-2}$, $S = 1.001$ and a droplet whose initial radius is $0.5\ \mu\mathrm{m}$, we calculate that it takes *ca.* 2000 s for the droplet to grow to a radius of $20\,\mu\mathrm{m}$, which is around the minimum size for the development of precipitation.

As can be seen from the discussion above, droplet growth by condensation is much too slow for the formation of rain drops of the size of millimetres. Instead, such large drops form by *collision-coalescence* (also known as accretion), whereby faster-falling large drops incorporate slower-falling small droplets. If droplets reach a size of *ca.* $10–20\,\mu\mathrm{m}$ by condensation, they start to settle under the influence of gravity, acquiring an appreciable fall velocity (*ca.* $1\,\mathrm{m\,s}^{-1}$). Consideration of the simple geometrical model shown in Figure 5.10(a) shows that collisional capture of small droplets by larger ones leads to a rate of growth given by

$$\frac{dm}{dt} = \pi(R+r)^2 w(V-v)f \tag{5.22}$$

where R and r are the radii of the large and small droplets respectively, V and v are their falling velocities, w is the total water content of the cloud (the mass of water per unit volume of air) and f is an aerodynamic correction factor. The factor f is ~ 1 for droplets with $R > 20\,\mu\mathrm{m}$, but decreases rapidly for smaller droplets. Coalescence relies upon the droplets having a non-zero relative velocity in the vertical direction, and is augmented by turbulent air flow. Under the assumptions that $R \gg r$ and $f = 1$, the growth rate is approximated as

$$\frac{dR}{dt} = \frac{wV}{4\rho_l} \tag{5.23}$$

The growth rates arising from both the coalescence and condensation mechanisms are illustrated in Figure 5.10(b). It is clear that condensation increases the size of smaller droplets and tends to homogenise the droplet

(a)   (b)

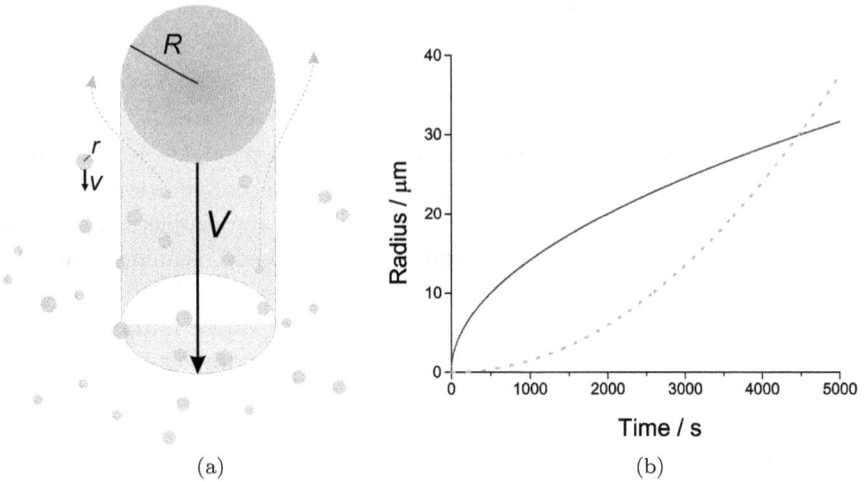

Figure 5.10   (a) Schematic of growth by collision-coalescence. (b) Comparison of growth rates for condensation (solid line) and collision-coalescence (dashed line). It was assumed that the collision efficiency was 0.05 for all radii and that the water content of the cloud was $1\,\mathrm{g\,m^{-3}}$.

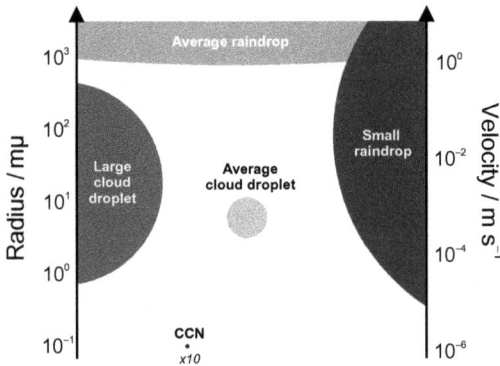

Figure 5.11   A size comparison between CCN and cloud droplets.

size distribution, whereas coalescence favours the largest droplets and leads to a separation of the droplet size distribution. Figure 5.11 summarises these discussions and shows the typical characteristics of particles involved in cloud and precipitation processes.

Finally, it should be noted that since the saturated vapour pressure over ice is less than that over liquid water, an ice crystal will grow more rapidly than a water droplet at the same temperature. Ice crystals will therefore grow at the expense of water droplets if both are present together. Ice particles can grow directly from the vapour phase by a process known as *riming* in which supercooled water droplets freeze upon falling ice particles, and may also form by aggregation of existing ice particles. Under the correct conditions crystals can grow to $100\,\mu$m or more in only a few minutes. Most precipitation at mid-Northern latitudes originates in this way, with the ice particles frequently melting before reaching the ground to produce rain rather than hail or snow. As in the case of liquid droplets, larger falling particles can collide with smaller ones and grow by coalescence; if the cloud depth is adequate, large droplets are produced in this very efficient process.

5.4 The Optical Properties of Aerosols and Clouds, and their Effects on Climate

The previous section highlighted the importance of aerosols in cloud formation. As introduced in Chapter 2, the interactions of both clouds and aerosols with atmospheric radiation play key roles in anthropogenic radiative forcing of the climate system. This is exemplified by Figure 2.20, which shows radiative forcing estimates between 1750 and 2011; it is clear that the forcing supplied by aerosols and clouds partly counteracts that of greenhouse gases (GHGs). The degree of radiative forcing is proportional both to the number of aerosol particles and to their ability to scatter and/or absorb radiation. The scattering efficiency is largest for accumulation mode particles with diameters comparable to the wavelength of visible light *ca.* 400–750 nm, and therefore all anthropogenic aerosols are good at scattering solar radiation. Such scattering preferentially prevents short-wavelength radiation from reaching the surface, leading to a decrease in the surface temperature. In addition, some aerosol particles, such as soot or black carbon, also absorb solar radiation, leading to warming of the atmosphere. While this absorbed radiation does not reach the surface and leads to surface cooling, the warmed atmosphere now re-emits the absorbed radiation at longer wavelengths. Some of this re-emitted radiation reaches the surface and tempers the decrease in surface temperature. Experiments find cooling at both the top of the atmosphere (TOA) and the surface, with much more cooling at the surface than at TOA due to aerosols. This is commonly known as the *direct* aerosol effect.

The warming of the atmosphere caused by optically-absorbing aerosols such as black carbon — see Section 2.2 — reduces the vertical temperature

Atmospheric Chemistry

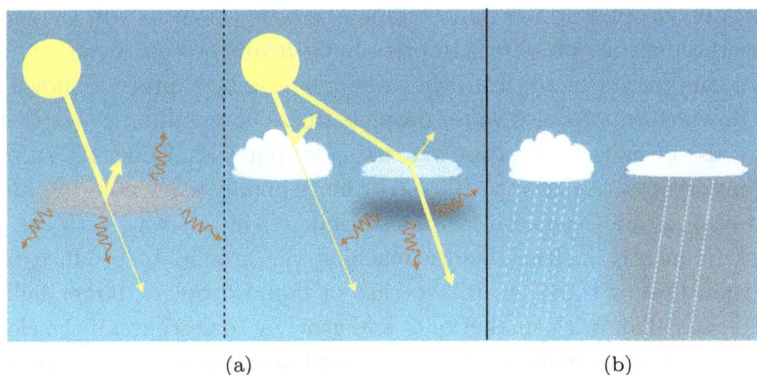

(a)   (b)

Figure 5.12   (a) The direct and the semi-direct aerosol effect on atmospheric radiation
and (b) aerosol–cloud interactions and their effect on atmospheric radiation.

gradient and leads to reduced vertical motion. In addition, absorption
within cloud droplets may cause the droplets to evaporate. The resulting
decreased cloud cover reduces the amount of reflected radiation, leading to
surface warming. This generates a feedback cycle, in which warming leads
to further warming, and for even a weakly absorbing layer, this *semi-direct*
effect can be larger than the direct effect (see Figure 5.12).

The fact that aerosol particles can act as CCN also affects cloud
properties and the radiation balance. An increase in the number density of CCN will increase the number density of cloud droplets. If the
liquid water content is fixed then this increase in number is accom-
panied by a decrease in the size of cloud droplets. The net effect is
to increase the surface area of the cloud, leading to a higher fraction
of solar radiation being reflected and scattered. Thus, a cloud formed
in a polluted environment will generally have a larger cross-sectional
area of particles, resulting in a higher cloud albedo; this aerosol–cloud
interaction is commonly known as the *indirect* aerosol effect or the
Twomey effect. An increase in aerosol particle number also leads to
changes in the microphysical structure of cloud. If the cloud consists of
smaller droplets, the precipitation rate is slowed because of the smaller
collision frequency. A reduced precipitation rate leads to less scavenging
of aerosol particles, causing particles to accumulate in the planetary
boundary layer (PBL). In contrast, in clean conditions, where only a
few CNN are present, clouds consists of fewer droplets and the cloud
droplet size distribution is rather broad, containing at least some large

droplets. Such a clean cloud will more readily form drizzle or rain via collision-coalescence than polluted cloud. Hence, a clean marine boundary layer tends to remain clean, whereas a polluted boundary layer tends to remain polluted.

Because aerosols are highly variable both spatially and temporally, forcing is much stronger near aerosol sources. However, the presence of aerosols can also change large-scale temperature and pressure gradients, resulting in marked circulation changes, for example winter warming after a volcanic eruption. Aerosol effects are difficult to quantify from observational data, as satellites are often redirected to cloud-free areas over oceans. Making such measurements over land is hampered by difficulties in accounting for variations in albedo, and data analysis relies on retrieval algorithms that rely on a number of assumptions. Such restricted ground-based measurements cannot monitor feedback effects directly, and in practice, quantifying aerosol forcing and feedback requires global climate modelling.

Clouds represent the biggest uncertainty in climate modelling. As in the case of aerosols, the reflection of solar radiation by clouds leads to reduced surface heating. In addition to this albedo effect, clouds also contribute to the greenhouse effect by absorbing and re-emitting the IR radiation emitted by the surface and lower atmosphere, i.e. clouds reduce the amount of terrestrial radiation transmitted to space, and thereby warm the Earth–atmosphere system. Whether the albedo effect or the greenhouse effect is dominant depends upon the latitude and thickness of the cloud and on its microphysical structure. For example, cirrus clouds, formed at altitudes in excess of 5 km, are semi-transparent in the visible region, and so have a small albedo effect. They cause a large greenhouse effect as a consequence of their high altitudes and low cloud-top temperatures; their emissivities are less than unity, and therefore less long λ radiation is emitted to space in their presence. In contrast, low level clouds reflect visible radiation efficiently, and their albedo effect dominates their greenhouse effect if they lie over a dark surface, such as an ocean. As these low clouds are located at low altitudes, the long λ radiation they emit is very efficiently absorbed and re-emitted by overlying layers of the atmosphere rather than being emitted directly to space; this means that their effect on outgoing long λ radiation is minimal.

Finally, it should be noted that bright surfaces such as ice or snow have enhanced reflectivity in the infrared (IR). Hence, IR radiation emitted from clouds back to the surface may by partially reflected from the surface and can be absorbed and re-emitted by the cloud. Such multiple cycles of reflection, absorption and re-emission can enhance the greenhouse effect of a low lying cloud as compared to a single reflection. The great importance of clouds in determining the radiation balance of the atmosphere can be appreciated by considering the fact that the *difference* in net radiation at the TOA between a cloudy and a cloud-free atmosphere is *ca.* $-47\,\mathrm{W\,m^{-2}}$ for short wavelengths and $+26\,\mathrm{W\,m^{-2}}$ for long wavelengths. The net effect of the clouds on the TOA radiation balance is therefore $-21\,\mathrm{W\,m^{-2}}$. This net cooling effect is some 5–6 times larger than the positive radiative forcing associated with the doubling of CO_2 levels ($3.7\,\mathrm{W\,m^{-2}}$) — see Section 2.4.2.

5.5 Reactions of Gases with Particles

A general overview of chemical kinetics within the atmosphere was presented in Chapter 1. We now consider the factors determining the kinetics of the uptake and/or reaction of gases with atmospheric particles. We will first consider the partitioning of soluble gases into aqueous aerosol or cloud droplets, before investigating some possible subsequent chemical reactions. Important classes of reaction include hydrolysis, oxidation and redox reactions. In particular, we will highlight the oxidation of sulphur dioxide to sulphuric acid:

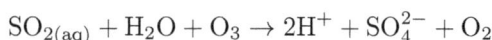

$$SO_{2(aq)} + H_2O + O_3 \rightarrow 2H^+ + SO_4^{2-} + O_2$$

which is an example of a reaction which results in both the chemical conversion of a trace gas and its removal in precipitation.

5.5.1 Partitioning of Soluble Gases into Droplets

The end products of gas-phase oxidation chemistry, which include species such as HNO_3, HCl, H_2O_2, and carboxylic acids, are often highly soluble and can be taken up readily into aqueous droplets. Gas solubility is quantified by *Henry's law*, which states that at a constant temperature, the equilibrium concentration c_A of a particular gas, A, dissolved in a given type and volume of liquid, is directly proportional to the partial pressure p_A of the gas.

$$c_A = Hp_A \tag{5.24}$$

where H is known as the Henry's Law constant and typically has units of $\mathrm{M\,atm^{-1}}$. Table 5.2 lists examples of Henry's law constants for a variety of gases dissolved in water at $298\,\mathrm{K}$. In cloud, the fraction of gaseous species

Table 5.2   Examples of Henry's law constants for a selection of trace gases dissolving in water at 298 K.

Gas	$H/\text{M atm}^{-1}$
O_2	1.3×10^{-3}
O_3	1.2×10^{-2}
SO_2	1.2
OH	30
HCHO	6.3×10^3
H_2O_2	10^5

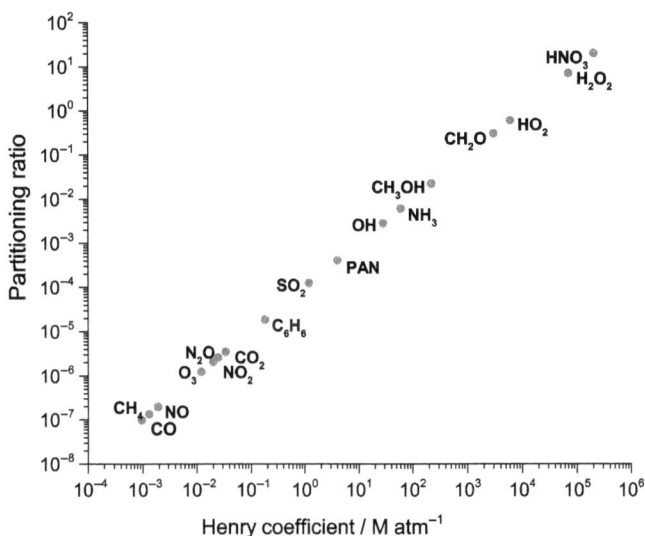

Figure 5.13   Gas–liquid partitioning for a selection of atmospheric trace species.

dissolved at equilibrium is $\sim L_c HRT$, where L_c is the liquid water content per unit volume of air. Once equilibrium is reached, the net uptake of gas into the droplet is zero. Figure 5.13 shows an example of the partitioning of various gases into a water droplet of radius $10\,\mu\text{m}$ at a particle density of $10^3\,\text{cm}^{-3}$ where the partitioning ratio is taken to be the ratio of the concentration in the droplet to that in the gas phase.

Acidic gases, HA, dissociate to a greater or lesser extent in water, i.e. $HA_{(aq)} \rightleftharpoons H^+_{(aq)} + A^-_{(aq)}$, leading to an effective increase in the solubility of the gas. The total amount of HA partitioned from the gas phase into the aqueous phase is pH-dependent, and is given by

$$[HA_{(aq)}] + [A^-_{(aq)}] = [HA_{(aq)}]\left(1 + \frac{K_a}{[H^+_{(aq)}]}\right)$$

where K_a is the acid equilibrium constant. An important example of such a situation is the solubility of SO_2 in tropospheric droplets. Here, the effective solubility of SO_2 is driven by the following two equilibria:

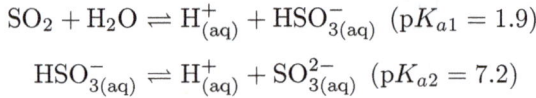

$$SO_2 + H_2O \rightleftharpoons H^+_{(aq)} + HSO^-_{3(aq)} \quad (pK_{a1} = 1.9)$$

$$HSO^-_{3(aq)} \rightleftharpoons H^+_{(aq)} + SO^{2-}_{3(aq)} \quad (pK_{a2} = 7.2)$$

The sum $[SO_{2(aq)}] + [HSO^-_{3(aq)}] + [SO^{2-}_{3(aq)}]$, constituting the total amount of S^{4+} in solution (denoted $[S(IV)]$), is then

$$[S(IV)] = Hp_{SO_2}\left(1 + \frac{K_{a1}}{[H^+_{(aq)}]} + \frac{K_{a1}K_{a2}}{[H^+_{(aq)}]^2}\right) \tag{5.25}$$

The absolute concentrations of $S(IV)$ species in a cloud droplet as a function of pH are shown in Figure 5.14 assuming an SO_2 volume mixing ratio of 1 ppbv. For pH < 6 it is clear that $HSO^-_{3(aq)}$ is the dominant $S(IV)$ species. Acid–base equilibria also increase the effective solubility of acidic and basic

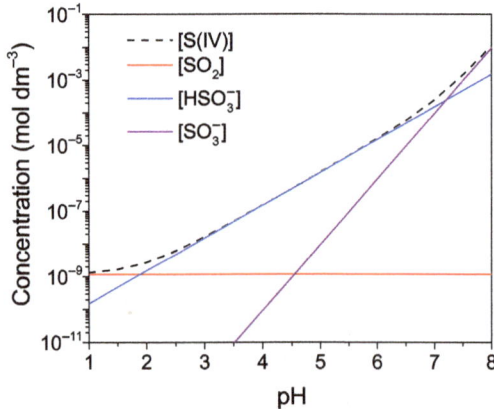

Figure 5.14 The absolute concentrations of $S(IV)$ species in a cloud droplet as a function of pH as determined by Equation (5.25).

Source: Adapted from J.H. Seinfeld and S.N. Pandis, *Atmospheric Chemistry and Physics: From Air Pollution to Climate Change*, Wiley (2006).

gases, with important examples including

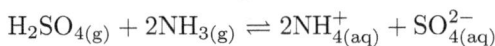

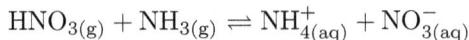

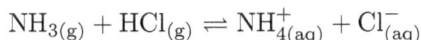

$$H_2SO_{4(g)} + 2NH_{3(g)} \rightleftharpoons 2NH_{4(aq)}^+ + SO_{4(aq)}^{2-}$$

$$HNO_{3(g)} + NH_{3(g)} \rightleftharpoons NH_{4(aq)}^+ + NO_{3(aq)}^-$$

$$NH_{3(g)} + HCl_{(g)} \rightleftharpoons NH_{4(aq)}^+ + Cl_{(aq)}^-$$

As noted previously, the first of these reactions is responsible for the predominant formation mechanism of sulphate aerosols.

5.5.2 Heterogeneous Reactions on Droplets

Attention now turns to the rates of heterogeneous reactions on droplet surfaces. From the kinetic theory of gases, it can be shown that the rate Z at which a gas-phase species A collides with a surface of unit area, is given by $Z = \bar{c}[A]/4$, where $\bar{c}$ is the mean speed of the gas molecules.[4] Introducing the *uptake coefficient*, γ, defined as the probability of removal from the gas phase per collision, allows the rate of removal at the droplet surface to be expressed as

$$\frac{d[A]}{dt} = -k_{het}[A] \tag{5.26}$$

where $k_{het} = (\gamma \bar{c} s)/4$, with s the surface area of the particles, equal to $4\pi r^2 n$ where n is the particle number density. Factors determining γ include gas-phase diffusion, accommodation (sticking to the surface), evaporation from the surface, solubility, and reaction, either at the gas-particle interface or in the condensed phase. These processes are shown schematically in Figure 5.15, and must all be considered in order to understand the overall rate of removal of matter from the gas phase. By analogy, and as discussed in Chapter 1, dry deposition can be treated as a first-order process in which the rate constant for removal, k_d, is simply given by $k_d = (\gamma \bar{c})/4$. Dry deposition rates are largest for HNO_3 and H_2O_2 because they stick very efficiently on most surfaces (high solubility). Dry deposition rates also depend strongly upon the type of surface, reflecting the type of intermolecular interaction between the surface and the trace gas. Dry deposition rates onto aqueous droplets are therefore smallest for non-polar and hydrophobic compounds.

---

[4] For a molecule at temperature, T, and having molar mass, M, $\bar{c} = (8RT/\pi M)^{1/2}$.

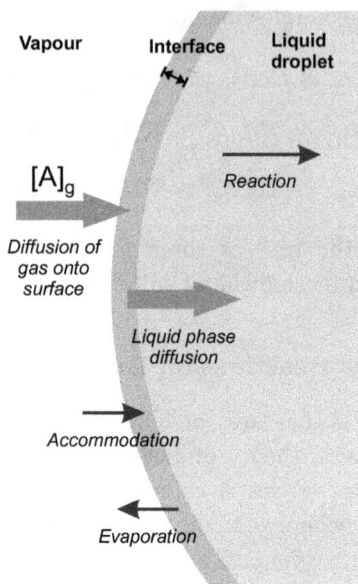

Figure 5.15   Factors determining the rate of heterogeneous reactions.

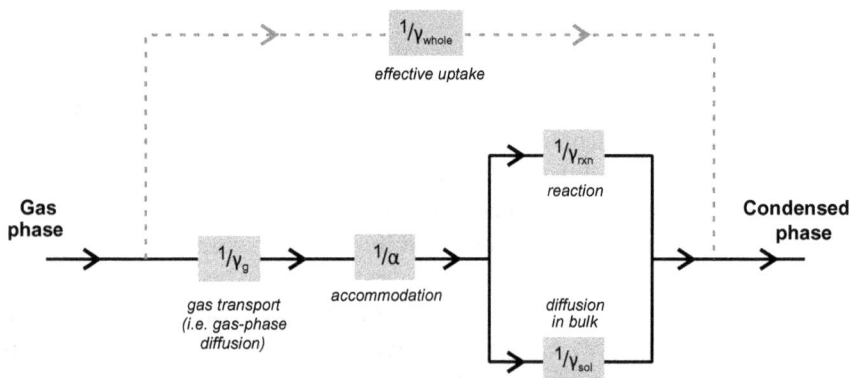

Figure 5.16   Generalised kinetic scheme (resistance model) for gas uptake.

The uptake of gases into aerosols and clouds can be treated in an analogous manner to the calculation of total resistances, R_{Tot}, in electrical circuits, as depicted in Figure 5.16.[5] The flux of a trace gas into the bulk of a

---

[5]See, for example, G.M. Nathanson *et al.*, *J. Phys. Chem.* **100** 13007 (1996).

liquid droplet is determined by the combined rate of a series of fundamental processes, each of which can be assigned a resistance. Multiple processes can be combined in series, $R_{\text{Tot}} = \sum_i R_i$ or parallel, $(1/R_{\text{Tot}}) = \sum_i (1/R_i)$. Here the resistances are just the inverse of the uptake coefficients, γ_i, and the overall net gas uptake coefficient is given by

$$\frac{1}{\gamma} = \frac{1}{\gamma_g} + \frac{1}{\alpha} + \frac{1}{\gamma_{\text{sol}} + \gamma_{rxn}} \tag{5.27}$$

where α is the mass accommodation coefficient — see below. Limits for the rate of gas uptake can then be identified as follows[6]:

(1) *Limited by diffusion*: When transfer to the condensed phase is fast, concentration gradients exist for the gas near to the surface of the particle, and gas-phase diffusion controls the uptake rate. This is often the case for large particles and in this limit the loss rate depends only on the size of the droplet.

(2) *Limited by mass accommodation*: The mass accommodation coefficient, α, determines the rate at which molecules cross the interface between the gas and condensed phase, and is defined as the ratio of the number of gas molecules absorbed by the condensed phase per second to the number of gas molecules colliding with the condensed phase per second. The magnitude of α determines the maximum rate of mass transfer; the net uptake of a particular gas, A, by the condensed phase is normally smaller due to re-evaporation. An important example of this limit is the reaction $N_2O_5 + H_2O_{(\text{surface})} \rightleftharpoons 2HNO_3$ which has the rate-determining step $N_2O_{5(g)} \rightleftharpoons N_2O_{5(aq)}$. In this case N_2O_5 dissolved in water disproportionates into NO_2^+ and NO_3^-, and NO_2^+ then reacts with water to yield HNO_3 and a proton.

(3) *Limited by solubility*: This limit is usually time-dependent in that the net uptake decreases with exposure time as the droplet saturates with trace gas. At equilibrium, the accommodation rate is equal to the rate of evaporation (note that the rate of evaporation increases as the concentration of A in the aerosol droplet increases). The larger the Henry's law coefficient, the slower the rate of evaporation. For example, considering the equilibrium $SO_{2(aq)} + H_2O \rightleftharpoons H^+ + HSO_3^-$, the solubility of SO_2 varies with pH, due to the different forms of sulphur (SO_2, HSO_3^-, and SO_3^{2-}) in equilibrium.

---

[6]For further details see, for example, T. Berkemeier *et al.*, *Atmos. Chem. Phys.* **13** 6663 (2013) and C.E. Kolb *et al.*, *Atmos. Chem. Phys.* **10**(66) 10561 (2010).

(4) *Limited by reaction*: Chemical reactions can occur at the surface of the
   droplet, or following diffusion into the bulk of the droplet. The former
   is most accurately termed a heterogeneous reaction, while the latter
   is described by bulk phase chemistry. Examples where gas uptake by
   the droplet is limited by chemical reaction at the surface include the
   reactions of Cl_2 with Br^- and OH with Cl^-, the latter proceeding
   via formation of a surface complex. Clearly, many reactions on solid
   surfaces, such as ice, mineral dust, or soot also fall into this limiting
   case. Examples of well-mixed bulk reaction systems include the reaction
   of O_3 with SO_2 under acidic conditions, the self-reaction of HO_2 in
   absence of transition metal ions, or reactions involving NO_2. Such
   reactions are typically not a major sink for the gas, but are important
   in terms of aerosol aging if they are the principle transformation of the
   condensed phase co-reagent. Experimentally, surface and bulk reactions
   can be differentiated by examining the dependence of the uptake
   coefficient on the concentration of the co-reagent within the droplet;
   if a linear dependence is observed, the reactive uptake is generally
   considered as progressing via a heterogeneous surface reaction.

As discussed above, oxidation of SO_2 in the troposphere occurs following
its transfer into aqueous droplets; the resultant sulphuric acid is the main
contributor to *acid rain*. Over typical pH ranges, dissolved SO_2 mostly
exists as HSO_3^- (see Figure 5.14) and the oxidants for converting S(IV)
into S(VI) are O_3 and H_2O_2. In the case of ozone, $[H^+]$ rises (pH falls)
as the reaction proceeds, and so the oxidation rate slows; this is a 'self-
inhibiting' reaction. In contrast, when H_2O_2 is the oxidant, the following
reactions occur:

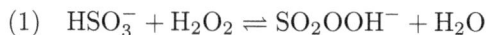

(1) $HSO_3^- + H_2O_2 \rightleftharpoons SO_2OOH^- + H_2O$

(2) $SO_2OOH^- + H^+ \rightarrow H_2SO_4$

The overall rate of reaction is directly proportional to the rate of reaction
step (2), which itself increases as $[H^+]$ rises. This compensates for the
drop in $[HSO_3^-]$, and so the overall conversion S(IV) $\rightarrow$ S(VI) by H_2O_2 is
nearly independent of pH over the range 2–5.5. Oxidation of SO_2 by H_2O_2
is therefore the dominant pathway at low pH, as shown in Figure 5.17,
although oxidation by ozone becomes significant at neutral pH. It should
also be noted that while the Henry's law constant for ozone is several orders
of magnitude smaller than for H_2O_2, the ozone levels in air are far larger.
The oxidation of S(IV) by O_2 is slow, but we note that it can be catalysed

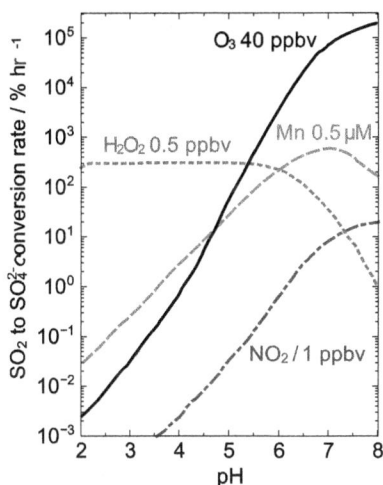

Figure 5.17   The oxidation of S(IV) to S(VI) in clouds as a function of pH for a selection
of oxidants. The relative reaction rates assume a liquid water content, L_c, of $0.3\,\mathrm{g\,m}^{-3}$.
Source: Adapted from *Chemistry of the Natural Atmosphere*, P. Warneck, ed., Academic
Press, San Diego (1988).

by traces of transition metals such as Fe^{3+} and Mn^{2+} present in cloud water
from mineral dust particles acting as CCN. If there is a sink for SO_2 in the
condensed phase, i.e. in the aerosol, then much more SO_2 is removed from
the gas phase, and so γ increases. In air, the uptake is diffusion controlled
for $r > 3.3\,\mu$m. As liquid water in clouds is mostly of this size or above, the
reaction is generally diffusion controlled in the atmosphere.

5.6 Impact of Aerosols on Health and Air Quality

In the previous chapter, the formation of elevated levels of ozone within
the PBL was highlighted as the result of the action of sunlight upon an
atmosphere containing significant quantities of NO_x and hydrocarbons.
Under these conditions ozone is a secondary pollutant, as are atmospheric
particles, whose presence is readily discerned by reduction in visibility. The
formation of this photochemical smog is aided by a stable atmospheric layer
under inversion which traps pollutants close to the ground. Formation of
smog results in problems associated with poor visibility, respiration, eye
irritation, building corrosion (NO_x gives rise to HNO_3), and crop/plant
damage. Weather conditions of low wind speed, temperature inversions,
and ground fog result in slow dispersion of species, leading to build-up

of concentrations of primary pollutants and poor air quality. Chemical transformations lead to the formation of secondary pollutants, and the local deposition of these pollutants damages surfaces.

An (in)famous example of smog formation is London smog, which was particularly severe in the winter of 1952. In this case SO_2, arising from uncontrolled coal burning, was taken up by fog droplets and oxidised to form sulphuric acid. Cold air from the English Channel settled over London and produced a temperature inversion which trapped pollution over a period of 5 days. The air pollution was so acute that some 4,000 people died of respiratory problems, and the smog was implicated in a further 8,000 subsequent deaths. This episode led to the first UK clean air acts, and legislation has now effectively removed the possibility of a recurrence throughout both the US and EU.

Emissions of fine particulate matter (PM) are regulated, particularly in the case of particles with a diameter less than 2.5 μm (PM2.5), as these particles penetrate easily into the lungs. At the most trivial level this uptake aggravates asthma and decreases lung capacity, but more seriously the smallest particles have enhanced toxicity that may cause cancer, and may also cause premature death in people with heart or lung disease. The World Health Organisation estimates that PM air pollution contributes to approximately 800,000 premature deaths each year, ranking it the 13th leading cause of mortality worldwide. Despite the introduction of air pollution strategies, small particles are still prevalent in significant concentrations in large cities and their long-term health effects are still not fully appreciated.

5.7 Summary

Aerosols are ubiquitous in the atmosphere and exist in a wide range of sizes (from a few nm to *ca.* 10 μm diameter), shapes and compositions. Tropospheric aerosols are composed of organic and inorganic substances originating from the direct emission of particles from the planetary surface and from condensation of gas-phase species. Aerosols are climate forcers, affecting the Earth's radiation balance by their intrinsic optical properties and by their intimate link to the formation of clouds. Aerosols are also important in that they provide surfaces and condensed phase environments within which chemical processing of trace gases may occur — they therefore affect both the health of our atmosphere and ourselves. Given the importance of such heterogeneous chemical processing, experimental and theoretical studies of the physical and chemical properties of multiphase systems such as aerosols and clouds continues to be a rich research area within atmospheric science.

5.8 Questions

5.8.1 Essay-style Questions

Q5.1: Discuss aerosol concentration and size distributions in urban air, highlighting factors which determine these quantities.

Q5.2: Briefly discuss the factors which determine the optical properties of aerosol particles.

Q5.3: (a) Explain why a very small drop of pure water has difficulty growing in the atmosphere.
(b) Explain how a speck of deliquescent dust aids droplet growth.
(c) Explain why the presence of such impurities is important for cloud formation.

Q5.4: Explain whether or not the scattering of solar radiation by clouds can be described by Rayleigh scattering.

Q5.5: Explain why significant (>1%) supersaturations with respect to liquid water are never observed in the atmosphere, but supersaturations with respect to ice are common.

Q5.6: Compare and contrast the radiative forcing of aerosols and GHGs.

5.8.2 Problems

P5.1: (a) What are the values of the saturation, RH, and supersaturation just above a pure water droplet of radius $0.05\,\mu m$ at $10°C$? Assume that $\gamma = 0.074\,N\,m^{-1}$ and $\rho = 3.4 \times 10^{28}$ molecules m^{-3}, respectively.
(b) If $10^{-19}\,kg$ of sodium chloride dissolves in the $0.05\,\mu m$ radius droplet, what is the saturation and RH just above the solution droplet? Assume that the density of the solution is $10^3\,kg\,m^{-3}$, that γ and ρ are the same as those for pure water, and that $T = 283\,K$. For NaCl, the van't Hoff factor is 2 because n

moles of NaCl dissolve into n moles of Na$^+$ and n moles of Cl$^-$ ions.

P5.2: During the homogeneous nucleation of water molecules to form a droplet of radius R, the net increase in free energy of the system, ΔG, due to the formation of the droplet can be expressed as:

$$\Delta G = 4\pi R^2 \gamma - \frac{4}{3}\pi R^3 n k_B T \ln \left(\frac{p}{p_0}\right)$$

where p and T are the vapour pressure and temperature of the system and p_0 is the saturation vapour pressure over a plane surface of water at temperature T. γ is the surface tension and n is the number of water molecules per unit volume of liquid.

(a) Sketch how ΔG varies as a function of R for both sub- and supersaturated conditions.

(b) Use the equation above to derive an expression for the critical radius, r^*, of a droplet that is in equilibrium with a given water vapour pressure p.

(c) Such an equilibrium is said to be unstable. Explain what is meant by this statement and *briefly* comment on the importance of homogeneous nucleation of pure water in cloud formation.

P5.3: Soluble aerosol particles in the atmosphere act as condensation nuclei that enable the formation of large cloud droplets. Within the Köhler theory the saturation, $S(r)$, over a salt water droplet of radius r is the product of two terms:

$$S(r) = \exp\left(\frac{a}{r}\right)\left(1 + \frac{b}{r^3}\right)^{-1}$$

where a and b are constant at a given temperature T.

(a) Discuss the significance of each of the terms in $S(r)$.

(b) Assume that $a \ll r$ and $b \ll r^3$ and thereby show that

$$S(r) \approx 1 + \frac{a}{r} - \frac{b}{r^3}$$

(c) Sketch $S(r)$ as a function of r.

(d) Derive an expression for the critical radius of r, r_c, and hence show that the value of S above which the particle will grow (and below which it will shrink) in saturated air, S_c, is given by:

$$S_c = 1 + \sqrt{\frac{4a^3}{27b}}$$

(e) For a droplet containing 10^{-17} kg of NaCl, the constants of a and b have values of $6 \times 10^{-3}\,\mu$m and $2 \times 10^{-5}\,\mu$m^3, respectively. Calculate the approximate critical radius of the droplet, and the minimum saturation ratio of the surrounding air for droplet growth.

P5.4: Consider two clouds in which 5×10^{-4} kg m^{-3} of water condenses. The CCN concentration in the each cloud is 100 and 1000 cm^{-3}, respectively.

(a) Assuming that the liquid water is equally distributed among the CCN, what is the droplet size in each cloud?
(b) What might the different concentrations of CCN indicate about the locations of the clouds?
(c) What are the consequences for the subsequent development of the clouds?

P5.5: Stokes' law states that the viscous force acting on a spherical drop of radius a and speed v is $6\pi\mu a v$ where μ is the dynamic viscosity for air.

(a) Show that Stokes terminal fall velocity, u, for a small water sphere in air is

$$u \approx \frac{2a^2 \rho_l g}{9\mu}$$

where ρ_l is the density of liquid water.
(b) Calculate the times taken for water drops of radii 1, 10, and $100\,\mu$m to fall a distance of 500 m in air at terminal velocity. Take $\mu = 1.7 \times 10^{-5}$ kg m^{-1} s^{-1}.

P5.6: A cloud droplet is activated and grows by condensation with rate $r\mathrm{d}r/\mathrm{d}t = C(S-1)$ where r is the droplet radius, $(S-1)$ is the supersaturation and C is a constant. The height $h(t)$ above the cloud base of a droplet growing through condensation can be expressed as

$$h(t) = wt - \left(\frac{2g\rho_l C(S-1)}{9\mu}\right) t^2$$

where ρ_l and μ are the density of liquid water and dynamic viscosity of air respectively, and can be assumed constant. Assume that the updraught velocity w and supersaturation also remain constant during droplet growth.

(a) Calculate the radius of a droplet under condensational growth 200 m above the cloud base, assuming a constant updraught velocity $w = 2\,\mathrm{m\,s^{-1}}$ and supersaturation $(S-1) = 5 \times 10^{-3}$. Take C to be $10^{-10}\,\mathrm{m^2\,s^{-1}}$ and μ to be $1.7 \times 10^{-5}\,\mathrm{kg\,m^{-1}\,s^{-1}}$.

(b) At an altitude of 200 m, the droplet concentration is $180\,\mathrm{cm^{-3}}$. Estimate the optical thickness at a wavelength of 550 nm for a 50 m horizontal path through cloud at this altitude. You may assume that the cloud droplets have no absorption at 550 nm, i.e. they only scatter radiation, and that the only absorber within the cloud is water vapour whose density is $10\,\mathrm{g\,m^{-3}}$. Take the mass absorption coefficient of water vapour at 550 nm to be $0.03\,\mathrm{m^2\,kg^{-1}}$.

P5.7: The viscous drag force acting on a falling spherical drop of radius $30 < a < 1000\,\mu\mathrm{m}$ is proportional to the *area* of the droplet (compare with Stokes Law in question (12.5)) and results in the droplet having a terminal velocity, $u \approx 8000a$.

(a) Use Equation (5.23) to calculate the time taken for a droplet to grow from a radius of 30 to $500\,\mu\mathrm{m}$ by the collisional coalescence in a cloud with a mean liquid water density of $1.2 \times 10^{-3}\,\mathrm{kg\,m^{-3}}$.

(b) How much longer is required to grow the droplet from 500 to $1000\,\mu\mathrm{m}$?

P5.8: (a) Consider the following loss processes for pernitric acid (HO_2NO_2) at an altitude of 20 km.

(i) $HO_2NO_2 + h\nu \rightarrow HO_2 + NO_2$ $J = 6 \times 10^{-6}\,\mathrm{s^{-1}}$
(ii) HO_2NO_2 loss on sulphate aerosol
$\gamma = 10^{-3}$, $s = 10^{-7}\,\mathrm{cm^2\,cm^{-3}}$
(iii) $HO_2NO_2 + OH \rightarrow$ products
$k = 1.3 \times 10^{-12} \exp(+380/T)\,\mathrm{cm^3\,molecule^{-1}\,s^{-1}}$

Calculate the lifetime of pernitric acid at 20 km assuming that the number density of air is 1.8×10^{18} molecules cm^{-3}, the temperature is 215 K and that the OH mixing ratio is 0.15 pptv.

(b) Sulphuric acid particles are found globally with increased abundance after volcanic eruptions. The background aerosol has a surface area density $s = 1 \times 10^{-8}$ cm^2 cm^{-3}. After a volcanic eruption, the aerosol density can increase by a factor of 50.

   (i) Assuming $\gamma = 0.1$, calculate the lifetime for the heterogeneous hydrolysis $N_2O_5 + H_2O \rightarrow 2HNO_3$ on sulphuric acid aerosols under background and volcanic conditions at an altitude of 20 km.

   (ii) How does the above heterogeneous reaction compare to the gas-phase reactions for N_2O_5 loss in the stratosphere? Assume the gas-phase reaction has a rate constant of $k = 1 \times 10^{-21}$ cm^3 molecule^{-1} s^{-1} and that water is present at 5 ppmv at 20 km.

P5.9: (a) Show that the partitioning ratio, α_A, for a species A between the gas and aqueous aerosol phases is given by

$$\alpha_A = \frac{p_A H_A v_L}{p_A H_A v_L + p_A/RT}$$

where H_A is the Henry's law coefficient for A and v_L is the volume of liquid aerosol per volume of air.

(b) With reference to Section 5.5, show that the total amount of S(IV) in solution is given by:

$$[S(IV)] = H p_{SO_2} \left(1 + \frac{K_{a1}}{[H^+_{(aq)}]} + \frac{K_{a1}K_{a2}}{[H^+_{(aq)}]^2} \right)$$

P5.10: What would be the pH of pure rainwater at an atmospheric pressure of 980 mbar? Assume that the atmosphere contains only N_2, O_2, and CO_2. Take the Henry's constant for CO_2 as 0.034 M atm^{-1}, the mixing ratio of CO_2 as 400 ppmv, and the equilibrium constant for the process $H_2CO_3 \leftrightarrow H^+ + HCO_3^-$ as 4.3×10^{-7}.

Appendix A

The Hydrostatic Equation

Section 1.2.1 introduced the concept of hydrostatic balance and presented the hydrostatic equation which quantifies the fall in atmospheric pressure, p, with increasing altitude, z. We now derive the hydrostatic equation with reference to Figure A.1. If we start at altitude z then on increasing altitude by an amount $\mathrm{d}z$, the mass of atmosphere above z decreases by an amount $\mathrm{d}m$, leading to a decrease in pressure given by $\mathrm{d}p = -g\mathrm{d}m/A$ where A is the cross-sectional area of the air mass. Now $\rho = m/V$ is the air density and so $\mathrm{d}m = -\rho A \mathrm{d}z$. Hence we arrive at the *hydrostatic equation*:

$$\mathrm{d}p = -\rho(z)g\mathrm{d}z \qquad (A.1)$$

This equation describes an atmosphere in static equilibrium, i.e. the gravitational force acting downwards is balanced by the pressure gradient force acting on the layer of air. For an ideal gas, $pV = nRT$, and so

$$\mathrm{d}p = -\frac{Mpg}{RT}\mathrm{d}z \qquad (A.2)$$

where M is the relative molar mass. For the simplest case of an isothermal atmosphere, i.e. at constant temperature, we can define the *scale height*, $H = RT/Mg$, and thus

$$\mathrm{d}p = -\frac{p}{H}\mathrm{d}z \qquad (A.3)$$

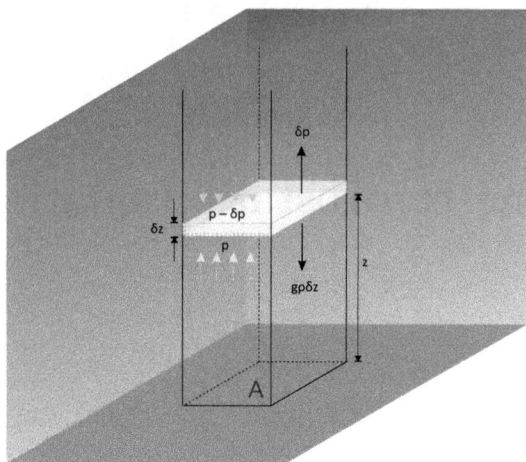

Figure A.1   The infinitesimal change of pressure with altitude.

Separating variables and integrating between the limits $z = 0$ to z and $p = p_0$ to $p(z)$,

$$\int_{p_0}^{p(z)} \frac{\mathrm{d}p}{p} = -\int_0^z \frac{\mathrm{d}z}{H} \tag{A.4}$$

then yields:

$$p(z) = p_0 \exp(-z/H) \tag{A.5}$$

Appendix B

The Saturated Adiabatic Lapse Rate

In Section 1.3, an expression was derived for the dry adiabatic lapse rate, Γ_d, of a hypothetical air parcel which does expansion work as it rises through the atmosphere. The magnitude of the (environmental) lapse rate in a real atmosphere is somewhat less than Γ_d and is a consequence of the fact that real air contains water vapour. As the parcel ascends water vapour may condense, thereby releasing heat into the air parcel and causing warming. Under these *saturated* conditions, the rising air parcel therefore experiences both expansion cooling and condensation warming with the net result that the parcel still cools on ascent, but at a slower rate. This new rate is called the *saturated adiabatic lapse rate*, Γ_s, a mathematical expression for which is derived below.

Consider a parcel of moist air that is lifted from the Earth's surface. The First Law of Thermodynamics states that the change in the internal energy of the parcel, dU, is given by

$$dU = dq + dw \qquad (B.1)$$

where dq is the heat supplied to the air parcel and dw is the work done on the air parcel. In this case, dq is not zero because water vapour may condense during the parcel's ascent. As shown in Chapter 1, enthalpy is defined as $H = U + pV$, and so $dH = dU + pdV + Vdp$. This expression can be combined with the First Law and the hydrostatic equation to give the following equation for dq:

$$dq = dH - Vdp = C_p dT - \frac{RT}{p}dp = C_p dT + gdz \qquad (B.2)$$

where C_p is the molar heat capacity at constant pressure. In this case, dq is simply the latent heat of condensation for water. At saturation, the water mixing ratio μ equals the saturation mixing ratio $\mu_s(T, p)$. If a mass $d\mu_s$ of water condenses as the parcel rises through the altitude dz then the amount of latent heat given to the parcel is then $dq = -Ld\mu_s$ where L is the latent heat of vapourisation *per unit mass* (also know as the *specific* enthalpy of vapourisation),[1] and so

$$C_p dT + g dz + L d\mu_s = 0 \qquad (B.3)$$

For the next step of this derivation it is necessary to express $d\mu_s$ in terms of dT and dz. To do this, we require the definition of the *saturation mixing ratio* as presented in Chapter 1,

$$\mu_s(T, p) = \frac{e_s(T)\epsilon}{p} \qquad (B.4)$$

which has units of grams (of water) per kilogram (of air), $g\,kg^{-1}$. In addition, we require the Clausius–Clapeyron equation which quantifies the saturation vapour pressure, e_s, of water as a function of temperature:

$$\frac{d \ln e_s}{dT} = \frac{\Delta H_{vap}}{RT^2} = \frac{L}{R_v T^2} \qquad (B.5)$$

where we note that we have converted between enthalpy of vapourisation per mole and enthalpy of vapourisation per kg; R_v is therefore the *specific* gas constant for water vapour. From these last two equations it follows that[2]

$$\frac{d\mu}{\mu_s} = \frac{de_s}{e_s} - \frac{dp}{p} = \frac{L}{R_v T^2} - \frac{dp}{p} \qquad (B.6)$$

From the hydrostatic equation, we have

$$dp = -\frac{Mpg}{RT} dz \qquad (B.7)$$

where p is the total pressure. Combining the last two equations leads to

$$\frac{d\mu}{\mu_s} = \frac{de_s}{e_s} - \frac{dp}{p} = \frac{L dT}{R_v T^2} + \frac{g dz}{RT} \qquad (B.8)$$

where c_p is the *specific* heat capacity of air at constant pressure. Hence

$$\left(c_p + \frac{L^2 \mu_s}{R_v T^2} \right) dT + \left(g + \frac{gL\mu_s}{RT} \right) dz = 0 \qquad (B.9)$$

---

[1]The parcel therefore undergoes a *non*-adiabatic change.

[2]Here we have used the relationship $d\ln x = x^{-1} dx$.

and so the change of the parcel's temperature with height for the saturated ascent can be written as

$$\Gamma_s = -\frac{dT}{dz} = \left(\frac{g}{c_p}\right) \frac{\left(1 + \frac{L\mu_S}{R_a T}\right)}{\left(1 + \frac{L^2 \mu_S}{c_p R_v T^2}\right)} \tag{B.10}$$

In the limit where μ_s is zero, the dry adiabatic laspe rate is recovered.

Answers to Numerical Problems

Chapter 1

1.2 0.025 K
1.3 (a) 0.465 kg m^{-3}, (b) 269.5 K, i.e. cooler than its environment
1.4 (a) 1.206 kg m^{-3} (moist air), (b) 1.22 kg m^{-3} (dry air), (c) 290.6 K,
(d) 10.56 g kg^{-1}
1.5 (d) 90.5 K; 2.6 kJ
1.6 (b) 346 K, (c) 2.1 atm
1.7 1.01 kJ K^{-1} kg^{-1}
1.9 (c) 4 K km^{-1}
1.10 (b) 531 s
1.11 (a) 13.3 years, (b) 5 × 10^5 molecules cm^{-3}, (d) 28 days
1.12 2.9%

Chapter 2

2.1 (c) 3 × 10^{-19} cm^3 molecule s^{-1}
2.2 2π; 125 cm^{-3}
2.3 (a) 55%
2.5 (b) 255.1 K, (c) (i) 243.5 K, (ii) 259.8 K
2.6 287.2 K
2.7 (c) 3.6 mbar

Chapter 3

3.1 (d) 5.1 × 10^{-4} s

3.3 (a) 1.5×10^{-3} s (20 km); 3 s (45 km), (b) 1.5×10^{-5} (20 km); 0.03 (45 km), (c) 2.2×10^7 (20 km); 2.8×10^4 s (45 km)

3.4 34.6 km

3.5 (d) $20 \, cm^{-3}$, (e) 9.4×10^7 molecules cm^{-3} (HO_2); 1.4×10^6 molecules $cm^{-3} \, s^{-1}$ (OH)

Chapter 4

4.1 (b) 1.4×10^8 molecules cm^{-3}, (c) 9.9 years

4.2 4.5×10^{11} kg year^{-1}

4.3 (a) 9×10^7 molecules $cm^{-3} \, s^{-1}$; 1.5×10^6 molecules cm^{-3}, (b) 1.85 hours, (c) 0.79

4.4 (b) (i) 0.99998, (ii) 4.7×10^6 molecules $cm^{-3} \, s^{-1}$

4.5 (b) 44 ppbv

4.6 (c) 0, (d) 9.9 pptv

Chapter 5

5.1 (a) 1.023; 102%; 2.3%, (b) 0.91; 91%

5.3 (e) 100 nm; 1.04

5.4 (a) 10.6 μm; 4.9 μm

5.5 (b) 4.27×10^4 s; 4270 s; 427 s

5.6 (a) 10.6 μm, (b) 6.4

5.7 (a) 1170 s, (b) 289 s

5.8 (a) 1.15×10^5 s, (b) (i) 1.95×10^5 s; 3.9×10^3 s, (ii) 1.1×10^8 s

5.10 5.6

Index

www.ingramcontent.com/pod-product-compliance
Lightning Source LLC
Chambersburg PA
CBHW060258220326

41598CB00027B/4153